SAS® Hash Objects
A Programmer's Guide

Michele M. Burlew

support.sas.com/bookstore

The correct bibliographic citation for this manual is as follows: Burlew, Michele M. 2012. *SAS® Hash Object Programming Made Easy*. Cary, NC: SAS Institute Inc.

SAS® Hash Object Programming Made Easy

ISBN 978-1-61290-098-8 (electronic book)
ISBN 978-1-60764-801-7

SAS Institute Inc., SAS Campus Drive, Cary, North Carolina 27513-2414

1st printing, September 2012

Contents

About This Book

Purpose

The goal of *SAS Hash Object Programming Made Easy* is to show you how hash objects in SAS DATA steps can be used to lookup data, combine data, and organize data. After reading the discussions and trying the examples, you should be able to start wisely incorporating hash object programming techniques in your applications.

Is This Book for You?

This book is aimed at all levels of SAS programmers who understand the basics of SAS DATA step programming. You should understand the purpose of a lookup table, the basics of combining data sets, and the reasons for working with keyed data.

The book frequently compares a hash object solution to a programming task with a solution that does not use hash objects. Most likely the solutions that do not use hash objects apply DATA step and PROC step techniques that you already know. Comparing the results of code that you already understand to the results of code that is new to you will help you understand when and how to use hash objects in the DATA step.

All levels of SAS programmers with minimal or no hash object experience should start at the beginning of the book. The first two chapters introduce you to the basics of hash objects and how to code them. These chapters also explain the terminology and data structures associated with hash objects.

Programmers who have a basic understanding of SAS hash objects can skip to Chapters 3, 4, and 5. These chapters present increasingly complex techniques and examples.

Chapter 6 describes how to manage SAS hash objects. This chapter is likely to be most useful to you if your experience with SAS hash objects is at an intermediate or higher level.

Overview of Chapters

This book covers SAS hash object programming concepts from the beginning level to the advanced level.

Preface: Tips on Determining Whether to Use a Hash Object. Lists tips to keep in mind as you learn how to use hash objects.

Chapter 1: Introduction. Introduces a simple hash object application and compares the application to one that does not use a hash object.

Chapter 2: Hash Object Terminology and Concepts. Defines the vocabulary and concepts associated with SAS hash object programming. This chapter explains the relationship between items in a hash object and variables in a DATA step.

Chapter 3: Basic Hash Object Applications. Introduces basic syntax of creating hash objects, referencing hash objects, and finding data in hash objects. The examples load data sets into hash objects; define key items and data items; and use hash objects as lookup tables or to summarize data. The chapter introduces hash iterator objects, which manage traversal of hash objects so that you can search through the items in the underlying hash objects.

Chapter 4: Creating Data Sets from Hash Objects and Updating Contents of Hash Objects. Explains how to create data sets from hash objects and how to modify the contents of hash objects during execution of a DATA step. The examples load data into and create data sets from hash objects. Several examples show how to summarize data using hash objects and hash iterator objects.

Chapter 5: Hash Objects with Multiple Sets of Data Items per Key Value. Explains the techniques and methods that process hash objects that work with multiple sets of data items per key value. The examples in this chapter are at a more advanced level than the earlier chapters. Examples include applications that sort and summarize data in hash objects and create data sets dynamically based on the contents of the hash objects.

Chapter 6: Managing Hash Objects. Explains the methods that can help you manage creation, reuse, and deletion of hash objects within a DATA step. Topics include how to estimate the size of hash objects. One example simulates BY-group processing by reusing a hash object.

Software Used to Develop the Book's Content

The examples in this book were developed using SAS 9.3 language and features. Although every effort has been made to include the latest information available at the time of printing, new features will be made available in later releases. Be sure to check out the SAS website for current updates and check the SAS Help and Documentation for enhancements and changes in new releases of SAS.

Data and Programs Used in This Book

The examples and data sets from this book are available for your use. These are available in a downloadable ZIP file from the author page for this book: support.sas.com/burlew.

All of the data sets used in the examples are fabricated data. Many data sets were generated using random number functions in SAS. The data sets have relatively simple structures and a minimal number of variables. You do not have to set any specific options to create the data sets or execute the programs. SAS macro language is not used in any of the examples.

Most of the output displays in the book are copies of PROC PRINT output sent to the RTF destination using the analysis style that is part of Base SAS.

Author Page

You can access the author page for this book at support.sas.com/burlew. This page includes several features that relate to this specific book, including more information about the book and author, book reviews, and book updates; book extras such as example code and data; and contact information for the author and SAS Press.

Additional Resources

SAS offers a rich variety of resources to help build your SAS skills and explore and apply the full power of SAS software. Whether you are in a professional or academic setting, we have learning products that can help you maximize your investment in SAS.

Bookstore	support.sas.com/publishing/
Training	support.sas.com/training/
Certification	support.sas.com/certify/
Higher Education Resources	support.sas.com/learn/
SAS OnDemand for Academics	support.sas.com/ondemand/
Knowledge Base	support.sas.com/resources/
Support	support.sas.com/techsup/
Learning Center	support.sas.com/learn/
Community	support.sas.com/community/
SAS Forums	communities.sas.com/index.jspa
User Community Wiki	www.sascommunity.org/wiki/Main_Page

Comments or Questions

If you have comments or questions about this book, you can contact the author through SAS as follows:

Mail: SAS Institute Inc.
SAS Press
Attn: Michele Burlew
SAS Campus Drive
Cary, NC 27513

Email: saspress@sas.com

Fax: (919) 677-4444

Please include the title of this book in your correspondence.

SAS Book Report

Receive up-to-date information about all new SAS publications via e-mail by subscribing to the SAS Book Report monthly eNewsletter. Visit support.sas.com/subscribe.

About The Author

Michele M. Burlew, president of Episystems, Inc., designs and programs SAS applications for data management, data analysis, report writing, and graphics for academic and corporate clients. A SAS user since 1980, she has expertise in many SAS products and operating systems. Burlew is the author of seven SAS Press books: *SAS Hash Object Programming Made Easy; Combining and Modifying SAS Data Sets: Examples, Second Edition; Output Delivery System: The Basics and Beyond (coauthor); SAS Guide to Report Writing: Examples, Second Edition, SAS Macro Programming Made Easy, Second Edition; Debugging SAS Programs: A Handbook of Tools and Techniques;* and *Reading External Data Files Using SAS: Examples Handbook.*

Learn more about the author by visiting her author page at support.sas.com/burlew. There you can download free chapters, access example code and data, read the latest reviews, get updates, and more.

Acknowledgments

I had two concerns when I started writing this book: Was there enough material to develop a full book and was the topic of hash objects too technical to interest many SAS programmers? As I dug into the project and with the help of many people, my concerns disappeared. For a SAS feature that has just a few methods and associated statements, a wealth of applications exists, from simple to complex.

I want to acknowledge the contributions of the following people who helped me write this book. To Jason Secosky and Janice Bloom, thank you for the conversations about hash objects and their applications that started me thinking about a proposal for a hash objects book. Thanks to the several reviewers at SAS who took the time to carefully review the draft: Mark Jordan, Al Kulik, Marjorie Lampton, Scott McElroy, Ted Meleky, Kent Reeve, Jason Secosky, Jan Squillace, and Kim Wilson. A special thanks to Scott for fielding several questions as I developed the examples!

Thank you to Russ Lavery for your thorough review of the first draft and your helpful comments. Thanks to the publications staff at SAS Press for making the book look good and read better: Candy Farrell, production specialist; Marchellina Waugh and Jennifer Dilley, graphic designers; Mary Beth Steinbach, copyeditor; Aimee Rodriguez and Stacey Hamilton, marketing; and Shelley Sessoms, acquisitions editor. Finally, a big thanks as always to SAS Press editor John West and SAS Press Editor-in-Chief Julie Platt for managing this book project and for making it possible for me to write books!

Preface

Tips on Determining Whether to Use a Hash Object

As you read through this book and try the examples, keep the following advice in mind.

Many factors determine whether you should use a hash object to solve a programming task. The two main considerations are:

- The ease with which you can code an appropriate hash object and to write the code that works with the hash object.
- Whether the processing speed improves if you use a hash object, and how important this savings is to your application.

Start simple. If you are new to programming with hash objects, it will take practice to make them a tool you readily reach for when you write your programs. The best approach is to start with simple applications that use hash objects as basic lookup tables. The more you do this, the more likely you will think of using hash objects in more complex ways, and the easier it will become to write them. Do not tackle hash objects that allow multiple sets of data items per key value until you're familiar with basic hash objects. Hash objects that allow multiple sets of data items per key value can be quite powerful. Once you comprehend how SAS processes a basic hash object, you will easily find reasons to try the features of hash objects that allow multiple sets of data items per key value.

Test your processing first with familiar tools. When you start a new program, write the code using PROCs and DATA steps in a way that is familiar to you when you solve your coding tasks. Then add simple hash objects to replace parts of your code. Check if your new code obtains the same results and whether processing times improve with hash objects. By comparing the results of your usual programming solutions with those that include hash objects, you will gain confidence in how to write and use hash objects appropriately.

Experiment to gain understanding of hash object programming syntax. Hash object programming can at first seem strange, and figuring out how and when the statements get compiled is best determined through experimenting with simple applications. While the statements that define a hash object are executable statements, in simpler applications they are more often treated as though they were declarative statements. Most examples in this book execute these statements once at the top of the DATA step, or in a block that executes once on the first iteration of the DATA step.

Remember processing constraints. As you code your DATA step that creates a hash object, consider how you will reference that hash object. Consider both memory usage and programming complexity when determining whether you should define a hash object.

Reducing the number of observations and restricting the data items loaded into the hash object to only those that are needed is a way to conserve memory. While it may seem counterintuitive, it may be more efficient to load your larger data set into the hash object, especially if it is your lookup data set. The actions of reading your smaller data set sequentially and looking up information in the hash object are likely to process more quickly than if you read your larger data set sequentially and look up information for each of its observations.

Chapter 1: An Overview of Hash Objects

Programming hash objects in SAS sounds like a geeky process. A quick review of the statements associated with hash objects shows that the statements look geeky too. Although the items that are associated with hash object programming seem to resemble statements or functions, they are not called statements or functions. Instead they are called *methods* and they're written in something called *dot notation*!

If you are reading this book, you probably already know how to write a DATA step. And you may have picked up this book to figure out if it's worth your time to learn a new way of processing data in the DATA step. You understand how SAS compiles and executes a DATA step. You know how to apply procedures like SORT, SQL, and FORMAT to structure your data the way you need to and to look up information efficiently.

If you are already a skilled SAS programmer and you know how to use SAS procedures effectively; why should you learn how to use hash objects? The answer is that they are an amazingly efficient tool for looking up information and joining information from data sets and tables.

Because you already know how SAS processes data, you can quickly learn how to write code that includes hash objects. The methods that define and work with hash objects are used in the DATA step in ways that are familiar to you.

This book shows you how to make hash objects fit into what you already know, and it often compares the hash solution to a solution that uses SAS language and procedures, one that you're likely already familiar with applying. This book introduces you to hash object programming and connects it to concepts you already understand.

If you write lots of SAS programs and use SAS procedures to structure and look up data, then it is worth it to take the time to learn how to use hash objects. Once you become familiar with the syntax, you will find many places where you can use a hash object. You are likely to save programming time and computer processing time when you incorporate hash object programming in your SAS programming toolkit. If you've already used hash objects some, then this book offers you some examples for techniques that you may not have tried or understood how to use.

The examples in this book vary in complexity and are generally organized in order of increasing complexity. By studying how an example uses a hash object, you will find ways to adapt hash objects programming techniques to your own applications.

What Are Hash Objects?

Technically, hash objects, interchangeably called hash tables in this book, are data structures that provide a way to efficiently search data. Hash object programming is a feature in many programming languages. A hash object is a type of array that a program accesses using keys. A hash object consists of key items and data items. The programming language applies a hash function that maps the keys to positions in the array.

When you define an array in SAS, SAS allocates a fixed amount of memory based on the number of elements you specify and the attributes of the variables that make up the array. A hash object is a more dynamic structure. It grows and expands as you add and remove items from the table following the instructions in your programs.

A SAS hash object exists only within the DATA step in which it creates the hash object. When the DATA step ends, SAS deletes the hash object.

Introducing a Simple Hash Object Application

In general, SAS language in a SAS DATA step follows a linear, logical process. A typical DATA step processes a SAS data set one observation at a time or it reads a raw data file one record at a time. SAS executes the statements one after the other. While your statements can direct processing to different locations within a DATA step, a DATA step typically starts at the top and proceeds sequentially statement-by-statement to the bottom of the DATA step.

A common SAS programming task is to look up information based on data in the observation currently being processed. Many methods exist in SAS to look up information. Some of these include:

- IF-THEN and SELECT statements
- referencing elements of arrays
- applying formats
- merging data sets
- joining tables

Some of these methods require more than just a single DATA step or PROC step to achieve the lookup.

For example, as shown in the first table in Output 1.1, say you have data set EMPHOURS with hours worked for a group of employees where each employee is identified in each observation solely by employee ID. To complete the employee report, you need the employee's full name. This data is stored in another data set, EMPLOYEES, as shown in the second table in Output 1.1 that contains both the employee ID and demographic information about the employee.

Output 1.1 PROC PRINT of EMPHOURS (all 6 observations) and EMPLOYEES (first 20 observations)

EMPHOURS		
Obs	**empid**	**emphours**
1	6XBIFI	38.5
2	WA4D7N	22.0
3	VPA9EF	43.0
4	TZ6OUB	11.5
5	L6KKHS	29.0
6	8TN7WL	38.0

EMPLOYEES							
Obs	**empid**	**empln**	**empfn**	**empmi**	**gender**	**startdate**	**emppaylevel**
1	6XBIFI	Ramirez	Danielle	N	F	04/21/1989	AIb
2	AWIUME	Thompson	Catherine	D	F	06/18/1986	PIIIa
3	06KH8Q	Chang	William	T	M	07/23/2002	PIIa
4	WA4D7N	Garcia	Breanna	X	F	08/20/1982	AIb
5	OOQT3Z	Jones	Brooke	E	F	08/28/1994	MIIa
6	1JU28B	Smith	Matthew	I	M	08/22/1982	TIIIb
7	V8OARE	Hall	Samuel	B	M	05/25/2010	PIb
8	1GTXQ2	Parker	Nathaniel	S	M	08/12/1996	PIc
9	VPA9EF	Baker	Cheyenne	C	F	02/24/1990	AIIa
10	0IP7L6	Hughes	Alexander	N	M	08/08/1991	TIIb
11	Q1A4SU	Sanchez	Nathaniel	W	M	08/13/1998	TIId
12	ANWFGX	Green	Tyler	I	M	12/04/1991	TIc
13	L1I8Y7	Edwards	Angelica	O	F	11/18/1991	MIIIa
14	TZ6OUB	White	Heather	T	F	01/27/1999	AIIIa
15	235TWE	King	Briana	M	F	10/08/1992	TIIc
16	XYOJC7	Scott	Mark	T	M	01/15/2002	TIIa
17	8TN7WL	Miller	Tyler	J	M	08/31/1998	AIIIc
18	US3DZP	Brown	Sarah	U	F	12/11/2000	TIb

EMPLOYEES							
Obs	**empid**	**empln**	**empfn**	**empmi**	**gender**	**startdate**	**emppaylevel**
19	ODBAIZ	Jones	Rachel	T	F	10/11/1999	PIIb
20	A4GJG4	Johnson	Angelica	Z	F	01/01/1994	AId

Typing a series of IF-THEN or SELECT statements to find an employee's name seems prohibitive, especially if your demographic lookup data set is large. Probably a merge in the DATA step or a PROC SQL join would work to obtain the names of the employees based on their employee IDs. If you do the merge, both your hours-worked data set and your demographic data set must be sorted or indexed by the employee ID prior to the DATA step that does the merge.

While PROC SQL does not require that you sort your tables before you have PROC SQL join them, the step processes more efficiently if your tables are sorted or indexed by the columns that do the join. However, when you use PROC SQL, if you need to do other processing of the table, like calculations, it can be more difficult to code those statements in SQL than if you wrote SAS language statements in a DATA step. If you add a second SELECT statement to your PROC SQL step or if you follow the PROC SQL step with a DATA step, you've now processed your data set twice.

Example 1.1 presents sample code that shows a DATA step merge solution.

Example 1.1 DATA Step Merge

```
proc sort data=emphours;
  by empid;
run;
proc sort data=mylib.employees;
  by empid;
run;
data empnames;
  merge emphours(in=inhours)
        mylib.employees(keep=empid empln empfn empmi in=inall);
    by empid;
  length empname $ 60;
  if inhours;
  if inall then empname=catx(' ',empfn,empmi,empln);
  else empname='** Not Found';
run;
```

With hash object programming, you can achieve both the lookup of employee demographic data and additional SAS language processing in one DATA step. Your DATA step starts by loading the demographic data set into a hash object.

To further streamline the lookup process, you can tell SAS to load only certain variables and observations into the hash object. For this simple example, you load the names of the employees in the hash object, and you tell SAS that the key to find data in the hash object is the employee ID.

Perhaps you know that your employee data set has observations for employees that have a specific job classification. You could tell SAS to load the hash object with the names of employees only from that subset of the employee population.

Since hash objects reside in memory, it's highly likely that the DATA step with the hash object runs more quickly than any of the solutions in the list above, and this is especially true if your data sets are large.

Example 1.2 presents a DATA step that uses a hash object to find employee names for administrative employees. The values of EMPPAYLEVEL for administrative employees start with "A". While the language in this step may look strange to you, it is really easy to understand and code once you have a little knowledge.

Example 1.2 DATA Step That Uses a Hash Object

```
data empnames;
  length empid $ 6 empln $ 30 empfn $ 25 empmi $ 1 empname $ 60;
  if _n_=1 then do;
   declare hash e(dataset: 'mylib.employees(where=(emppaylevel=:"A")');❶ ❷
   e.definekey('empid'); ❸
   e.definedata('empln','empfn','empmi'); ❹
   e.definedone();
   call missing(empln,empfn,empmi); ❺
  end;
  set emphours; ❻
  drop rc;
  rc=e.find();❼
  if rc=0 then empname=catx(' ',empfn,empmi,empln); ❽
  else empname='** Not Found'; ❾
run;
```

The step starts with an IF-THEN block that executes only on the first iteration. This code defines hash object E and loads it with demographic information from data set EMPLOYEES. Having been defined on the first iteration of the DATA step, hash object E remains in memory throughout execution of the DATA step so that you can quickly retrieve from it the demographic data you need for each observation in the EMPHOURS data set.

The statements that follow the IF-THEN block execute once for every observation in EMPHOURS. The statement with the FIND method retrieves information from hash object E based on the values of EMPID. The IF-THEN statement that follows it executes when SAS finds a match in hash object E. The ELSE statement executes when SAS does not find a match in hash object E.

Specifically, the statements in the IF _N_=1 block that executes on the first iteration of the DATA step do the following:

❶ Create hash object called E.

❷ Load into hash object E the observations in MYLIB.EMPLOYEES for administrative employees (EMPPAYLEVEL=:"A").

❸ Identify variable EMPID as the key to find data in hash object E.

❹ Load only variables EMPLN, EMPFN, and EMPMI from MYLIB.EMPLOYEES into hash object E as data.

❺ Initialize to missing the DATA step variables with the same names as the items that SAS loads into hash object E.

The statements outside of the IF _N_=1 block that execute on each iteration of the DATA step do the following:

❻ Read each observation from data set EMPHOURS.

❼ Look for a match in hash object E for the value of EMPID in the EMPHOUR observation currently being processed.

❽ When SAS finds a match in hash object E, concatenate the information retrieved from hash object E into variable EMPNAME.

❾ When SAS does not find a match in hash object E, assign informative text to the variable EMPNAME.

The LENGTH statement at the beginning of the DATA step adds to the Program Data Vector (PDV) the three variables whose values are retrieved from hash object E. SAS outputs these three variables, the employee ID (EMPID), and the new variable EMPNAME to data set EMPNAMES.

Example 1.2, which is the DATA step with the hash object, executes the same way as a DATA step without a hash object. You could think of the block of code that defined hash object E as similar to how you might define an array that provides lookup values. Yet, hash object programming provides much more functionality than arrays. In this simple DATA step in Example 1.2, you can see that in one statement the DATA step looks for a match in hash object E and it retrieves the values of three data items when SAS finds a match. If you used arrays, your code would likely require multiple arrays, or possibly multi-dimensional arrays.

Chapter 2: Hash Object Terminology and Concepts

This chapter defines the main terms and explains the general concepts associated with hash objects. These terms and concepts are applied throughout this book and in SAS documentation.

Hash object programming in the DATA step uses elements associated with object-oriented programming. This aspect of SAS has its own set of terminology that is different from the SAS language elements with which you are already familiar.

Hash objects are used only in DATA steps. You use SAS language statements to work with hash objects. To access the elements of a hash object, you use method calls that are invoked in SAS language statements.

What Is a SAS Hash Object?

A SAS hash object is a data structure that resides in memory. It is not a data set, an array, a format, or any other SAS structure you might have already encountered.

A hash object consists of key items and data items. SAS determines how to structure a hash object so that it can efficiently retrieve data from the hash object based on information it stores with the key values.

SAS stores hash objects in memory. Hash objects are not stored on disk. Storage and retrieval of information from a hash object is fast since the object resides in memory. Efficiency and ease of access to information is a primary reason for incorporating hash object programming in your applications.

A hash object can be used only in a DATA step, and it exists only during execution of the DATA step that defines it. Unless you explicitly output the contents of a hash object to a SAS data set, its contents disappear when the DATA step ends.

You can load a data set into a hash object. You can fill it with data as your DATA step executes, and then output its contents to a data set before the DATA step ends. You can use a hash object like you would an array or format when you want to look up data. However, its features are much more powerful than an array or format, and you can more easily program complex lookups with hash objects than you can with arrays and formats.

Defining Terms Associated with Hash Objects

A hash object is a predefined *component object*. When you work with hash objects in the DATA step, you write SAS statements that create and manipulate these component objects.

Component objects are data elements that consist of *attributes*, *methods*, and *operators*.

An *attribute* is a property of the object.

A *method* is an operation that an object can perform.

An *operator* in the context of working with hash objects in SAS *instantiates* a hash object.

Instantiation is the creation of a component object such as a hash object.

SAS 9.3 has five types of components objects:

- hash objects and hash iterator objects
- Java objects
- logger objects and appender objects

This book focuses only on hash objects and hash iterator objects.

You can store and retrieve data from a *hash object*.

A *hash iterator object* is always associated with a hash object. The methods that a hash iterator object uses manage traversal of the associated hash object in a forward or backward direction.

Example 2.1 repeats the code in Example 1.2 from Chapter 1. In the context of this DATA step and highlighted in Example 2.1:

- The *component object* is the hash object named "E".
- Hash object E was *instantiated* with the DECLARE statement.
- Hash object E performs four *methods*: DEFINEKEY, DEFINEDATA, DEFINEDONE, and FIND.

Example 2.1 DATA Step That Defines a Hash Object

```
data empnames;
  length empln $ 30 empfn $ 25 empmi $ 1 empname $ 60;
  if _n_=1 then do;
    declare hash e(dataset: 'mylib.employees(where=(emppaylevel=:"A")');
    e.definekey('empid');
    e.definedata('empln','empfn','empmi');
    e.definedone();
    call missing(empln,empfn,empmi);
  end;

  set emphours;
  drop rc;

  rc=e.find();
  if rc=0 then empname=catx(' ',empfn,empmi,empln);
  else empname='** Not Found';
run;
```

The example did not examine any attributes of hash object E. Only two attributes exist in the SAS 9.3 implementation of hash objects. One attribute, ITEM_SIZE, obtains the size of an item in the hash object. The other attribute, NUM_ITEMS, obtains the number of items in the hash object.

Writing Code That Works with Hash Objects

The methods in the DATA step in Example 2.1 are written in *dot notation*. In dot notation, you place the object's name to the left of the dot, or period, and the method name or attribute name to the right of the dot.

When working with methods, you may want to add options called *argument tags*. Argument tags are enclosed in parentheses following the method name. If your method call does not include any argument tags, you must still include a set of empty parentheses to terminate the call.

The syntax to call a method is:

```
rc=object.method(<argument_tag-1: value-1<,…argument_tag-n: value-n>>);
```

SAS assigns a numeric code after a method executes that indicates whether the method executed successfully. The syntax statement above assigns that value to variable RC. You can alternatively write your code to omit assigning the return code to a DATA step variable. In this situation, if the method executes in error, SAS writes an error message in the SAS log.

For most hash object methods, SAS assigns a return code of 0 to indicate success of the method and a non-zero return code to indicate failure. Always check SAS documentation for how SAS sets return codes for the version of SAS that you are using.

Following good programming practices, a program should examine return codes and then test these values so that different processing actions can be taken. This is especially true when you do not fully know your data sets. Since most examples in this book use simple, easy-to-understand data sets, many programs do not test the return codes.

Attribute calls do not have any argument tags or terminating sets of parentheses. The syntax to call an attribute is:

```
variable-name=object.attribute;
```

An attribute returns the value of a property of the hash object. You can assign this value to a DATA step variable, as shown in the above syntax statement, or it can be included in a SAS expression.

Understanding How SAS Stores Hash Objects in Memory

A big advantage of working with hash objects is that SAS dynamically allocates memory as it needs it. You do not have to determine the size of your hash object before you can use it. For example, you can reuse the same code that defines your hash object even if the next time you use it, you have many more observations to load into it.

This flexibility is different than if you worked with an array of SAS variables. When you define an array with the ARRAY statement, you must specify the number of elements in your array. If the number of elements changes the next time you use your DATA step, you must update your ARRAY statement, or possibly maintain additional code like macro programs that could update this for you.

The amount of memory your SAS session has available determines how big your hash object can be. The amount of space that the hash objects in the examples in this chapter take is trivial. On the other hand, it is possible that you may not have enough memory to create a hash object from a data set that has millions of observations and hundreds of data items.

You can make a rough estimate of the amount of space your hash object might take by multiplying the number of observations by the observation length, or more precisely by using the ITEM_SIZE attribute. Chapter 6 describes the ITEM_SIZE attribute. However, even if your hash object fits into memory, other processing that you're doing within the DATA step can affect memory usage.

As you code your DATA step that creates a hash object, consider how you will reference that hash object. Consider both memory usage and programming complexity when determining whether you should define a hash object.

Reducing the number of observations and restricting the data items loaded into the hash object to only those that the program needs is a way to conserve memory. While it may seem counterintuitive, it may be more efficient to load your larger data set into the hash object, especially if it is your lookup data set. The action of reading your smaller data set sequentially and looking up information in a large hash object is likely to process more quickly than if you read your larger data set sequentially and look up information for each of its observations in a small hash object.

Another way to conserve memory if your DATA step is complex and you have multiple hash objects is to delete hash objects after they are no longer needed. You can also empty out a hash object and refill it. Chapter 6 shows how these actions can be performed.

Understanding How Long Hash Objects Persist

A constraint on working with a hash object is that it exists only during execution of the DATA step that creates it. If you create a hash object in a DATA step and then run a second DATA step in which you want to use the same hash object, you must recreate it in the second DATA step.

You can create a SAS data set from a hash object with the OUTPUT method, but this does not also save the hash object itself for use later in your SAS session.

Specifying the Contents of Hash Objects

The items that make up a hash object are *keys* and *data.*

When you create a hash object, you must also define at least one key. Keys make the linkage between the DATA step and the hash object. All hash objects store and retrieve data based on the values of keys.

While not required, you will often want your hash object to store data items. When you store or retrieve information from your hash object by key values, SAS stores or retrieves data associated with the key values from the hash object.

In the Example 2.1 DATA step, the call to the DEFINEKEY method specifies one key, EMPID, in hash object E:

```
e.definekey('empid');
```

The call to the DEFINEDATA method specifies EMPLN, EMPFN, and EMPMI as data items:

```
e.definedata('empln','empfn','empmi');
```

When you instantiate a hash object with the DECLARE statement and specify key items and data items with the DEFINEKEY and DEFINEDATA methods, you simply define the object and its characteristics. This definition does not pass information to the SAS compiler that the key items and data items are variables. You must add SAS language statements like LENGTH and ATTRIB to define the key items and data items as DATA step variables. This manual action is required even when you fill your hash object with data from a data set, which was how hash object E in Example 2.1 was created.

The LENGTH and ATTRIB statements add variables to the Program Data Vector (PDV) whose names are the same as names of your hash object's key items and data items. Then when SAS retrieves a key value from a hash object, SAS places the value in the corresponding DATA step variable in the PDV so that you can work with the key value in SAS language statements in your DATA step. Similarly, data item values associated with the key value are retrieved from the hash object and placed in the DATA step variables whose names are the same as the names of the data items.

At the time you create a hash object, you have control over the hash object's structure and how you want your DATA step to process it. Additionally, when you fill your hash object with a data set, you can apply many familiar SAS data set options to the data set. Example 2.1 demonstrates that you can fill hash object E with observations where EMPPAYLEVEL starts with "A", which identifies observations for administrative employees.

```
declare hash e(dataset: 'mylib.employees(where=(emppaylevel=:"A")');
```

Modifying a Hash Object As a DATA Step Executes, and Creating a Data Set from a Hash Object

Your DATA step can fill a hash object with data, remove data from a hash object, and modify data in a hash object as the step executes. Methods ADD and REF place data in a hash object. The REMOVE method removes entries from a hash object. The REPLACE method modifies entries in a hash object. At the conclusion of the DATA step, or at any appropriate point during execution of the DATA step, the OUTPUT method can write the contents of a hash object to a SAS data set.

By default, a hash object has one set of data items per key value. You can override this default when you need to work with multiple sets of data items per key value. Chapter 5 describes the methods that manage this kind of hash object.

Examples in Chapters 4, 5, and 6 show how you can create data sets from hash objects. A useful option in this situation is the ORDERED: "YES" argument tag on the DECLARE statement. This argument tag tells SAS to return or output items that you store in a hash object by the object's key values. When you output such a hash object to a data set, SAS writes the observations in order by the key values because you defined the hash object in that way. When SAS accesses this data set later, it recognizes that this data set is in sorted order. Most likely you will save computer resources when you use this feature because you will not have to follow your DATA step with a PROC SORT step.

The DATA step in Example 2.2 shows how you can fill a hash object with data as the DATA step executes, and then create a data set from the contents of the hash object. The ADD method adds data from the PDV to hash object YEARS only when processing observations for employees who have worked for ten years or more. The last action of the DATA step writes the contents of YEARS to data set EMPS10PLUS.

Example 2.2 Filling a Hash Object As a DATA Step Executes, and Creating a Data Set from a Hash Object

```
data _null_;
  attrib empid length=$6
         empln length=$30
         empfn length=$25
         empmi length=$1
         startdate length=8 format=mmddyy10.
         empyears length=8 label='Number of years worked'
         yearscat length=$10 label='Years worked category';

  if _n_=1 then do;
    declare hash years(ordered: 'yes');
    years.definekey('yearscat','empln','empfn','empmi');
    years.definedata('empid','empln','empfn','empmi','empyears','yearscat');
    years.definedone();
    call missing(empid,empln,empfn,empmi,startdate,empyears,yearscat);
  end;

  set employees end=eof;

  empyears=ceil(('31dec2012'd-startdate)/365.25);

  if empyears ge 10;

  if 10 le empyears lt 15 then yearscat='10+ years';
  else if 15 le empyears lt 20 then yearscat='15+ years';
  else if 20 le empyears lt 25 then yearscat='20+ years';
  else if empyears ge 25 then yearscat='25+ years';
```

```
      rc=years.add();

      if eof then rc=years.output(dataset: 'emps10plus');

   run;
```

Because Example 2.2 defined hash object YEARS with the ORDERED: 'YES' argument tag, SAS retrieves data from hash object YEARS in ascending order by the values of the four key variables: YEARSCAT, EMPLN, EMPFN, and EMPMI. After SAS reads the last observation from EMPLOYEES, it creates data set EMPS10PLUS and writes observations to EMPS10PLUS in order by YEARSCAT, EMPLN, EMPFN, and EMPMI. The variables specified in the DEFINEDATA method are the only variables that SAS writes to data set EMPS10PLUS.

Output 2.1 shows the first 10 observations of EMPS10PLUS.

Output 2.1 PROC PRINT of EMPS10PLUS (first 10 observations)

Obs	empid	empln	empfn	empmi	empyears	yearscat
1	5F9U8L	Adams	Alexander	X	13	10+ years
2	YY5RXY	Adams	Christopher	Z	13	10+ years
3	YOOPWK	Adams	Joshua	I	10	10+ years
4	4A25WB	Adams	Lucas	M	10	10+ years
5	KNMW87	Adams	Michael	G	12	10+ years
6	P6L8NB	Adams	Nathaniel	H	12	10+ years
7	JSSKV0	Adams	Nathaniel	N	14	10+ years
8	ACU122	Adams	Thomas	C	10	10+ years
9	Y4RIRR	Adams	Tiffany	W	13	10+ years
10	FIQ6AU	Alexander	Jenna	I	13	10+ years

Initializing Variables in a DATA Step That Contains a Hash Object

When you load a data set into a hash object, your SAS language code must define and reference the DATA step variables that form the key items and data items in your hash object. The ATTRIB statement in Example 2.2 satisfies this requirement. If you do not include these actions, the SAS compiler issues notes that your variables are uninitialized and your DATA step may not compile and execute correctly.

When you define a hash object, you only instantiate the object into which you load a data set. The SAS compiler does not recognize at the time of instantiation that a link exists between the hash object's key items and data items to the variables in the data set that SAS will load into the hash object. In other words, even though the key items and data items look like DATA step variables, the compiler does not automatically add them to the PDV when the DEFINEKEY or DEFINEDATA methods execute.

Further, if any of your data items is a character variable and you do not define it as such, your DATA step may not compile and execute correctly. SAS may generate type mismatch errors in this situation because by default SAS defines a variable as numeric unless you explicitly define its type as character.

The LENGTH statement in the DATA step in Example 2.1 defines and initializes the key and data variables that are defined as key and data items in the hash object. Example 2.1 defines the data item variables, the key item variable EMPID, and the new DATA step variable EMPNAME with the LENGTH statement:

```
length empid $ 6 empln $ 30 empfn $ 25 empmi $ 1 empname $ 60;
```

Example 2.1 also initializes to missing the three variables whose names are the same as the hash object's data items:

```
call missing(empln,empfn,empmi);
```

The CALL MISSING statement is executable so be careful where you place it in your DATA step. In its use discussed here, place the statement so that it executes once *before* your DATA step processes the data set that the SET statement reads. If you place it outside of this IF _N_=1 block, the CALL MISSING statement may set to missing the variables SAS reads from your data set.

By contrast, Example 2.2 does not load a hash object with data from a data set. If you remove the ATTRIB statement and CALL MISSING statement from Example 2.2, the DATA step still executes correctly because SAS does not add entries to hash object YEARS until *after* the SET statement executes. When SAS begins execution of the DATA step, SAS adds the variables in data set EMPLOYEES to the PDV. The definitions for these variables are also applied to the key and data items in hash object YEARS that have the same names. Example 2.2 and others like it in this book include the ATTRIB and CALL MISSING statements so that a consistent programming style is followed in writing DATA steps that use hash objects.

Illustrating How the Program Data Vector Connects DATA Step Variables and Hash Object Items

As mentioned in the previous section, your DATA step must define the key items and data items in your hash object when it loads a hash object with a data set. A common way to do this is to start the DATA step with a LENGTH statement or an ATTRIB statement that defines these items. SAS also defines variables when statements that read data execute. These data reading statements include the INPUT, SET, and MERGE statements. When you use data reading statements, you may need to add code to control how these statements affect the contents of the PDV.

All of these actions define the variables in the PDV. When SAS compiles a DATA step, it allocates space in the PDV for each variable in the input data set and for each variable created by DATA step statements, including LENGTH and ATTRIB. SAS automatically adds the two variables _N_ and _ERROR_ to the PDV. These automatic variables track the processing status of the DATA step.

For the DATA step to communicate with the hash object, the key items and data items must be defined in the PDV. When SAS retrieves a data item from the hash object, it copies the item's value to the space allocated in the PDV for the variable with the same name as the data item. When SAS stores a DATA step variable's value in the same-named data item in a hash object, it copies the value that exists in the PDV for that variable.

Even though your DATA step might not return data from the hash object or write data to the hash object, you must still define variables with the same names and attributes as your hash object key items. For example, the CHECK method only looks for the presence of a key value in the hash object and it does not return any data. When SAS calls the CHECK method, it still uses the PDV for communication between the DATA step and the hash object.

The following six DATA steps in Examples 2.3 through 2.8 illustrate some of the concepts of how SAS handles DATA step variables and hash object items when SAS loads a data set into a hash object. All of the DATA steps execute a PUT _ALL_ statement, which lists in the SAS log the entire contents of the PDV for the observation currently being processed.

Examples 2.3 through 2.8 process three data sets: CONFROOMS, ROOMSCHEDULE, and ROOMSCHEDULE2. Data set CONFROOMS contains descriptive information for several conference rooms. Data sets ROOMSCHEDULE and ROOMSCHEDULE2 contain the conference room schedule for August 13, 2013. Variable ROOMID is present and identically defined in the three data sets. The value for ROOMID in the third observation of ROOMSCHEDULE and ROOMSCHEDULE2 is the only difference between these two data sets.

Examples 2.3 through 2.8 each load data set CONFROOMS into hash object CR. Hash object CR has a single key, ROOMID.

Output 2.2 lists the contents of CONFROOMS.

Output 2.2 PROC PRINT of CONFROOMS

Obs	roomid	roomno	floor	building	capacity
1	A0210	10	2	Anderson	50
2	A0120	20	1	Anderson	75
3	B0B05	5	B	Baylor	100
4	B1004	4	10	Baylor	15
5	B0212	12	2	Baylor	30
6	C0P01	1	P	Cummings	150

Output 2.3 lists the contents of ROOMSCHEDULE.

Output 2.3 PROC PRINT of ROOMSCHEDULE

Obs	meetingdate	meetingtime	roomid
1	08/13/2013	8:30	C0P01
2	08/13/2013	11:30	B1004
3	08/13/2013	1:15	A0120
4	08/13/2013	2:00	B1004

Output 2.4 lists the contents of ROOMSCHEDULE2. The only difference between it and data set ROOMSCHEDULE is the value of ROOMID in the third observation.

Output 2.4 PROC PRINT of ROOMSCHEDULE2

Obs	meetingdate	meetingtime	roomid
1	08/13/2013	8:30	C0P01
2	08/13/2013	11:30	B1004
3	08/13/2013	1:15	A0122
4	08/13/2013	2:00	B1004

Omitting Statements That Define the Hash Key Items and Data Items

Example 2.3 shows that if you do not define the hash key items and data items when you load a data set into a hash object, SAS might not compile and execute the DATA step.

The code in Example 2.3 tells SAS to create hash object CR, and load CR with data found in data set CONFROOMS. It defines one key, ROOMID, and four data items: ROOMNO, FLOOR, BUILDING, and CAPACITY. No statements in the DATA step define these items as variables. Neither an ATTRIB statement nor a LENGTH statement is present. Variable ROOMID is the only variable in common between data set ROOMSCHEDULE that the SET statement reads and data set CONFROOMS.

Example 2.3 reads data set ROOMSCHEDULE, and looks for entries in CR where the value for ROOMID in CR is equal to the value in ROOMSCHEDULE.

Example 2.3 DATA Step That Omits Statements That Define Hash Key Items and Data Items

```
data pdvck1;
  if _n_=1 then do;
    declare hash cr(dataset: 'confrooms');
    cr.definekey('roomid');
    cr.definedata('roomno','floor','building','capacity');
    cr.definedone();
  end;

  set roomschedule;
  rc=cr.find();

  put _all_;
run;
```

The SAS log in Output 2.5 shows that SAS was not able to execute Example 2.3. The first error message states that data item ROOMNO is undeclared. The step stops after examining this first data item. It does not let you know whether it found the remaining data items undeclared (which they are). For the DATA step to compile and execute, at least one of the following three statement types must be included in the DATA step:

- an ATTRIB statement or LENGTH statement that explicitly defines the key and data items
- a data reading statement where SAS reads variables with the same names and attributes as the key and data items
- assignment statements prior to the hash definition that define variables with the same names and attributes as the key and data items

Output 2.5 SAS Log Excerpt for Example 2.3

```
992  data pdvck1;
993    if _n_=1 then do;
994      declare hash cr(dataset: 'confrooms');
995      cr.definekey('roomid');
996      cr.definedata('roomno','floor','building','capacity');
997      cr.definedone();
998    end;
999
1000    set roomschedule;
1001    rc=cr.find();
1002
1003    put _all_;
1004  run;
```

```
ERROR: Undeclared data symbol roomno for hash object at line 997
column 5.
ERROR: DATA STEP Component Object failure. Aborted during the
EXECUTION phase.
NOTE: The SAS System stopped processing this step because of errors.
WARNING: The data set WORK.PDVCK1 may be incomplete. When this step
was stopped there were 0 observations and 4 variables.
```

Including Statements That Define the Hash Key Items and Data Items

Example 2.4 also loads data set CONFROOMS into hash object CR. This DATA step corrects the errors in Example 2.3 by defining the four data items with an ATTRIB statement. It is not necessary to include ROOMID in the ATTRIB statement since the SET statement references a data set that includes variable ROOMID.

The DATA step includes a CALL MISSING statement that initializes the four data items: ROOMNO, FLOOR, BUILDING, and CAPACITY. These four variables are not present in data set ROOMSCHEDULE so without the CALL MISSING statement SAS cannot initialize these variables. If you omit the CALL MISSING statement, SAS generates notes advising that these four variables are uninitialized.

Example 2.4 DATA Step That Defines a Hash Object's Key Items and Data Items

```
data pdvck2;
  attrib roomno   length=8
         floor    length=$2
         building length=$20
         capacity length=8;
  if _n_=1 then do;
    declare hash cr(dataset: 'confrooms');
    cr.definekey('roomid');
    cr.definedata('roomno','floor','building','capacity');
    cr.definedone();

    call missing(roomno,floor,building,capacity);
  end;
  set roomschedule;
  rc=cr.find();

  put _all_;
run;
```

The SAS log in Output 2.6 shows that Example 2.4 executes without errors. For each observation in ROOMSCHEDULE, the FIND method looks in hash object CR for the observation's value of ROOMID. If SAS finds the value of ROOMID in CR, it copies the values for ROOMNO, FLOOR, BUILDING, and CAPACITY in CR to the same-named variables in the PDV. All values for ROOMID in ROOMSCHEDULE are present in data set CONFROOMS. The PUT _ALL_ statement output shows that SAS assigned the correct data values from hash object CR to DATA step variables ROOMNO, FLOOR, BUILDING, and CAPACITY.

Data set ROOMSCHEDULE contains four observations and SAS outputs that number of observations to data set PDVCK2.

Output 2.6 SAS Log Excerpt for Example 2.4

```
NOTE: There were 6 observations read from the data set WORK.CONFROOMS.
roomno=1 floor=P building=Cummings capacity=150 meetingdate=08/13/2013
meetingtime=8:30 roomid=C0P01 rc=0 _ERROR_=0 _N_=1
roomno=4 floor=10 building=Baylor capacity=15 meetingdate=08/13/2013
meetingtime=11:30 roomid=B1004 rc=0 _ERROR_=0 _N_=2
roomno=20 floor=1 building=Anderson capacity=75 meetingdate=08/13/2013
meetingtime=1:15 roomid=A0120 rc=0 _ERROR_=0 _N_=3
roomno=4 floor=10 building=Baylor capacity=15 meetingdate=08/13/2013
meetingtime=2:00 roomid=B1004 rc=0 _ERROR_=0 _N_=4
NOTE: There were 4 observations read from the data set WORK.ROOMSCHEDULE.
NOTE: The data set WORK.PDVCK2 has 4 observations and 8 variables.
```

Reviewing the PDV after Looking for a Key Not Found in a Hash Object

The DATA step in Example 2.5 is identical to that in Example 2.4 except that it reads ROOMSCHEDULE2 instead of ROOMSCHEDULE. The value ROOMID='A01222' in the third observation of ROOMSCHEDULE2 is not present in data set CONFROOMS, and thus the key value of 'A01222' is also not present in hash object CR.

This DATA step shows that SAS sets to missing the variables in the PDV that have the same names as the hash data items when it reads an observation from ROOMSCHEDULE2. It does not retain the values from the previous observation.

Example 2.5 DATA Step That Looks for a Key That Is Not Present in a Hash Object

```
data pdvck3;
  attrib roomno   length=8
         floor    length=$2
         building length=$20
         capacity length=8;
  if _n_=1 then do;
    declare hash cr(dataset: 'confrooms');
    cr.definekey('roomid');
    cr.definedata('roomno','floor','building','capacity');
    cr.definedone();

    call missing(roomno,floor,building,capacity);
  end;
  set roomschedule2;
  rc=cr.find();

  put _all_;
run;
```

Because ROOMID='A01222' is not in hash object CR, SAS cannot find any data to retrieve from CR to insert into ROOMNO, FLOOR, BUILDING, and CAPACITY in the PDV. The PUT _ALL_ statement output for this observation shows that the values of these four variables are missing in the third (_N_=3) observation. The output for this statement is **bold** in Output 2.7. Note the non-zero return code (RC=160038) following execution of the FIND method for this third observation.

Output 2.7 SAS Log Excerpt for Example 2.5

```
NOTE: There were 6 observations read from the data set WORK.CONFROOMS.
roomno=1 floor=P building=Cummings capacity=150 meetingdate=08/13/2013
meetingtime=8:30 roomid=C0P01 rc=0 _ERROR_=0 _N_=1
roomno=4 floor=10 building=Baylor capacity=15 meetingdate=08/13/2013
meetingtime=11:30 roomid=B1004 rc=0 _ERROR_=0 _N_=2
roomno=. floor=  building=  capacity=. meetingdate=08/13/2013
meetingtime=1:15 roomid=A0122 rc=160038 _ERROR_=0 _N_=3
roomno=4 floor=10 building=Baylor capacity=15 meetingdate=08/13/2013
meetingtime=2:00 roomid=B1004 rc=0 _ERROR_=0 _N_=4
NOTE: There were 4 observations read from the data set WORK.ROOMSCHEDULE2.
NOTE: The data set WORK.PDVCK3 has 4 observations and 8 variables.
```

Looking for a Key Value in a Hash Object without Returning Any Data

Example 2.6 verifies the presence of a key in a hash object without retrieving any data. It uses the CHECK method, which does not return any data items to the DATA step as the FIND method does. Examples 2.4 and 2.5 demonstrated the FIND method.

Example 2.6 looks for two specific entries in hash object CR, and CR does not have any data items. The CHECK method looks for the presence of a value of key item ROOMID in hash object CR. Example 2.6 demonstrates that even if you do not return data, you need to define the variables that serve as key items. The ATTRIB statement at the start of the DATA step defines variable ROOMID. The DATA step references ROOMID only in the ATTRIB statement and in the DEFINEKEY method call. It does not read with a SET statement a data set that contains this variable nor does it contain any assignment statements that give it a value.

The DATA step invokes the CHECK method twice, each time specifying a literal key value for SAS to look for in hash object CR. The CHECK method returns only a code to inform you whether it found the key in the hash object. The value SAS returns when it finds the key is 0. When SAS does not find the key value, it returns a non-zero value. Example 2.6 assigns the return code value to variable RC.

Example 2.6 DATA Step That Does Not Assign a Value to the Variable That Also Serves As a Key Item in a Hash Object

```
data pdvck4;
  attrib roomid length=$5;

  declare hash cr(dataset: 'confrooms');
  cr.definekey('roomid');
  cr.definedone();

  call missing(roomid);

  rc=cr.check(key: 'C0P01');
  put 'CHECK 1: ' _all_;
  rc=cr.check(key: 'D0110');
  put 'CHECK 2: ' _all_;
run;
```

Output 2.8 shows that the value for ROOMID remains missing throughout execution of the DATA step. SAS does not deposit in the PDV the literal values passed to the hash object through variable ROOMID.

The lookup in CR for literal 'C0P01' is successful. SAS assigns RC=0, and SAS does not assign a value to ROOMID. The lookup for literal 'D0110' is not successful; the value for RC is non-zero; and the value of ROOMID is missing.

Output 2.8 SAS Log Excerpt for Example 2.6

```
NOTE: There were 6 observations read from the data set WORK.CONFROOMS.
CHECK 1: roomid=  rc=0 _ERROR_=0 _N_=1
CHECK 2: roomid=  rc=160038 _ERROR_=0 _N_=1
NOTE: The data set WORK.PDVCK4 has 1 observations and 2 variables.
```

Defining Variables by Using An IF Statement and SET Statement

Examples 2.3 through 2.6 process only a few variables. When you have many items to load into your hash object, it can be tedious to write an ATTRIB statement or a LENGTH statement that defines the items. A way to add many variables to the PDV is to start the DATA step with an IF statement whose condition is false and its action is the SET statement. While SAS does not execute the action on the false IF statement, it does compile the SET statement. The compilation of the SET statement adds to the PDV all of the variables in the data set named on the SET statement.

```
if _n_=0 then set confrooms;
```

Although this technique can save you some coding effort, it does require that you initialize the PDV after you attempt to retrieve data from the hash object. In contrast, Example 2.5 does not use this technique. When Example 2.5 does not find a key in the hash object, the DATA step variables with the same names as the hash object items have missing values. If you use the IF _N_=0 THEN SET *data-set-name* technique described in this section and SAS does not find a key in the hash object, the DATA step variables with the same names as the hash object items retain the data values of the previous observation.

Example 2.7 modifies Example 2.5 by adding an IF _N_=0 THEN SET *data-set-name* statement at the beginning of the program and removing both the ATTRIB and CALL MISSING statements. It processes ROOMSCHEDULE2. The third observation in ROOMSCHEDULE2 where ROOMID='A01222' is not present in CONFROOMS.

The IF _N_=0 condition is always false so the SET statement never executes. However, the presence of the SET statement on the IF statement causes SAS to add the variables in CONFROOMS to the PDV.

Example 2.7 Defining Variables by Using an IF Statement and SET Statement

```
data pdvck5;
  if _n_=0 then set confrooms;

  if _n_=1 then do;
    declare hash cr(dataset: 'confrooms');
    cr.definekey('roomid');
    cr.definedata('roomno','floor','building','capacity');
    cr.definedone();

  end;
  set roomschedule2;
  rc=cr.find();

  put _all_;
run;
```

Because ROOMID='A01222' is not in hash object CR, SAS has no data to retrieve from CR for this value of ROOMID. The data that was retrieved from CR when SAS processed the second observation remains in the PDV when SAS processes the third observation in ROOMSCHEDULE2. The PUT _ALL_ statement output for this third observation (_N_=3) shows that the values of these four variables are the same as the values in the second observation (_N_=2).

The output for the third observation is **bold** in Output 2.9. Note that the return code from the FIND method for this third observation is not zero.

Output 2.9 SAS Log Excerpt for Example 2.7

```
NOTE: There were 6 observations read from the data set WORK.CONFROOMS.
roomid=C0P01 roomno=1 floor=P building=Cummings capacity=150
meetingdate=08/13/2013 meetingtime=8:30 rc=0 _ERROR_=0 _N_=1
roomid=B1004 roomno=4 floor=10 building=Baylor capacity=15
meetingdate=08/13/2013 meetingtime=11:30 rc=0 _ERROR_=0 _N_=2
roomid=A0122 roomno=4 floor=10 building=Baylor capacity=15
meetingdate=08/13/2013 meetingtime=1:15 rc=160038 _ERROR_=0 _N_=3
```

```
roomid=B1004 roomno=4 floor=10 building=Baylor capacity=15
meetingdate=08/13/2013 meetingtime=2:00 rc=0 _ERROR_=0 _N_=4
NOTE: There were 4 observations read from the data set WORK.ROOMSCHEDULE2.
NOTE: The data set WORK.PDVCK5 has 4 observations and 8 variables.
```

A way to correct the error in Example 2.7 is to set the variables that were also defined as data items in the hash object to missing when the FIND method does not find the key in the hash object. You would not set the key item to missing since the key item is also a variable in data set ROOMSCHEDULE2.

Example 2.8 executes the CALL MISSING statement when the return code from the FIND method indicates that a match for the key value is not present in hash object CR. The CALL MISSING statement sets to missing the four variables also defined as data items in CR.

Example 2.8 Setting Variables to Missing When a Key Value Is Not Found in a Hash Object

```
data pdvck6;
  if _n_=0 then set confrooms;

  if _n_=1 then do;
    declare hash cr(dataset: 'confrooms');
    cr.definekey('roomid');
    cr.definedata('roomno','floor','building','capacity');
    cr.definedone();

  end;
  set roomschedule2;
  rc=cr.find();
  if rc ne 0 then call missing(roomno,floor,building,capacity);

  put _all_;
run;
```

Output 2.10 shows the SAS log from Example 2.8 with the output for the third observation in **bold**.

Output 2.10 SAS Log Excerpt for Example 2.8

```
roomid=C0P01 roomno=1 floor=P building=Cummings capacity=150
meetingdate=08/13/2013 meetingtime=8:30 rc=0 _ERROR_=0 _N_=1
roomid=B1004 roomno=4 floor=10 building=Baylor capacity=15
meetingdate=08/13/2013meetingtime=11:30 rc=0 _ERROR_=0 _N_=2
roomid=A0122 roomno=. floor= building= capacity=.
meetingdate=08/13/2013 meetingtime=1:15 rc=160038 _ERROR_=0 _N_=3
roomid=B1004 roomno=4 floor=10 building=Baylor capacity=15
meetingdate=08/13/2013 meetingtime=2:00 rc=0 _ERROR_=0 _N_=4
```

Chapter 3: Basic Hash Object Applications

This chapter presents simpler applications of hash objects and hash iterator objects. These examples show you how to define, load data into, and look up information in these component objects.

Using a Hash Object As a Lookup Table

One of the simplest ways to use a hash object is as a place to store your lookup data that is keyed by one variable whose values are unique. You start by defining a hash object in a DATA step and loading the hash object with data from your lookup data set or other source. Then, as you process observations in a second data set, you look up information in the hash object based on a key value derived from the second data set.

For example, suppose you have a data set with codes that identify medical procedures for a group of patients for which you need the text description for the procedures performed on the patients. Another data set stores procedure codes and text descriptions for all possible procedures. This reference data set is your lookup data set, and your DATA step loads its data into a hash object. The procedure code variable, which is common both to your patient data set and your procedure information data set, is the key item that links the two data sources. As your DATA step processes the patient data set sequentially, your code finds the text description for the procedures that each patient received by looking up each procedure code in the hash object and retrieving its associated descriptive data from the hash object.

Using a hash object as a lookup table is similar to working with an array or with a user-defined format. Some advantages of a hash object over an array or a user-defined format are:

- You do not have to specify the number of items you will store in the hash object as you would when you define the number of elements in an array.
- You do not have to precede the DATA step with a PROC FORMAT step in which your VALUE statement must specify all values that you want to look up.
- You can retrieve one or more items when you use a hash object compared to only one item when referencing an array element or format value.

Defining a Hash Object

Before you can use a hash object in a DATA step, you must declare and instantiate the hash object. This can be done two different ways. Examples in this book use the second method.

1. Define the hash object with the DECLARE statement, and instantiate the hash object with the _NEW_operator.

```
declare hash myhash;
myhash = _new_ hash();
```

2. Define and instantiate the hash object with the DECLARE statement.

```
declare hash myhash();
```

Table 3.1 lists the six optional argument tags on the DECLARE statement. Examples throughout this book make use of these argument tags. The DATASET, HASHEXP, ORDERED, and SUMINC argument tags are independent from each other. Specifications for the DUPLICATE and MULTIDATA argument tags can have an effect on each other.

Table 3.1 DECLARE Statement Argument Tags

DECLARE Statement Argument Tag	Description
`dataset: 'dataset_name <(datasetoption)>'`	Specifies the name of the data set to load into the hash object. SAS data set options can be applied to the data set. For example, variables can be renamed with the RENAME= option, and observations can be selected with the WHERE= option.

`duplicate: 'option'` where `'option'` can be
`'replace' | 'r'`
stores the last duplicate key record.
`'error' | 'e'`
reports an error to the log if a duplicate key is found.
	Specifies whether to ignore duplicate keys when loading a data set into a hash object. Neither option is a default. Chapter 5 contains several examples that use this option and demonstrate its interaction with the MULTIDATA argument tag.

`hashexp: n`	Specifies the hash object's internal table size where *n* is a power of 2 (2^n). Chapter 6 describes how to use this argument tag.

`multidata: 'option'` where `'option'` can be
`'YES' | 'Y'`
allows multiple data items for each key.
`'NO' | 'N'`
allows only one data item for each key.
	Specifies whether multiple sets of data items are allowed per key. The default option is no. Chapter 5 contains several examples that use MULTIDATA: "YES" and demonstrate its interaction with the DUPLICATE argument tag.

(continued)

Table 3.1 *(continued)*

DECLARE Statement Argument Tag	Description
`ordered: 'option'` where `'option'` can be `'ascending' | 'a' | 'YES' | 'Y'` SAS returns data in ascending key-value order. Specifying `'ascending'` is the same as specifying `'yes'`. `'descending' | 'd'` SAS returns data in descending key-value order. `'NO' | 'N'` SAS returns data in some undefined order.	Specifies whether or how SAS returns data in key-value order or if you use the hash object OUTPUT method. Many examples in this book use the ORDERED argument tag.
`suminc: 'variable-name'`	Maintains a summary count of hash object keys in a DATA step variable. Example 4.11 presents an example that uses SUMINC.

Finding Key Values in a Hash Object

The two methods whose sole function is to look for key values in a hash object are CHECK and FIND. When you use your hash object as a lookup table, you will most likely use one of these two methods. Other methods exist that find data in a hash object, but they have additional functionality, such as when you traverse a hash object in key value order; when you add, remove, or modify data in the hash object; or when you have multiple sets of data items for a key value.

The CHECK method determines whether a key value exists in the hash object. The FIND method also determines whether a key value exists in the hash object. If data items are defined in the hash object, the FIND method returns those data items' values to the corresponding DATA step variables. The syntax for these two methods is:

```
rc=object.CHECK(<KEY: keyvalue-1,..., KEY: keyvalue-n>);

rc=object.FIND(<KEY: keyvalue-1,..., KEY: keyvalue-n>);
```

A hash object key item can be either a character item or a numeric item. When you have more than one key item, these items can be a mix of character items and numeric items.

The KEY argument tag in CHECK and FIND is optional. If the DATA step variable has the same name as the key item, you probably do not need to specify the KEY argument tag. If the key value that you want to look up is a literal value or if the DATA step variable has a different name than the key item, you need to specify the KEY argument tag.

You can specify the key value that you want SAS to find in your hash object in any of the following four ways.

1. The simplest way to link to the hash object is if your DATA step variable that has the key value has the same name and attributes as the key item in your hash object. In that situation, you can write the method without specifying the KEY argument tag.

```
rc=h.find();
```

2. When the value you want to look up is in a DATA step variable that has a different name than the name of your key item, you link to the key item in the hash object by specifying the KEY argument tag followed by the name of the DATA step variable. Do not enclose the variable name in quotation marks. If your hash object has multiple key items, you must specify the KEY argument tag for each key item in the same order that you listed the key items on the DEFINEKEY method. This holds true even if a key item has the same name and attributes as a DATA step variable.

```
rc=states.check(key: stateabbrev);
rc=sales.find(key: id, key: transactiondt);
```

3. When your lookup value is a literal, you use the KEY argument tag followed by the literal value. If your hash object has multiple key items, you must specify the KEY argument tag for each key item in the order that you listed the key items on the DEFINEKEY method. This holds true even if a key item has the same name and attributes as a DATA step variable.

```
rc=zips.check(key: 55101);

rc=states.find(key: 'MN');

rc=locs.find(key: stateabbrev, key: 53);
```

4. When you need to construct the key value from a combination of variable values and literal values, or if you need to convert a value from character to numeric, you use the KEY argument tag followed by the elements that construct your lookup value. If your hash object has multiple key items, you must specify the KEY argument tag for each key item in the order that you listed the key items on the DEFINEKEY method. This holds true even if a key item has the same name and attributes as a DATA step variable.

```
rc=area.find(key: cats(statepostal, put(zipcode,z5.)));

rc=ages.find(key: put(zipcode,z5.), key: agecat);
```

Defining the Key Structure in a Hash Object

The default structure of a hash object allows only one set of data items per key value. However, you can add the MULTIDATA argument tag to the DECLARE statement so that your hash object can have more than one set of data items per key value. The examples in this chapter show you how to use the CHECK and FIND methods only under the condition of one set of data items per key value.

Chapter 5 describes how to work with hash objects that allow multiple sets of data items per key value. Methods exist that find all the entries in your hash objects when your key values are not unique.

Understanding How the FIND and CHECK Methods Alter the Values of DATA Step Variables and Hash Object Data Items

When you retrieve data from a hash object with the FIND method, SAS overwrites the values of the same-named variables in the Program Data Vector (PDV) for the observation that it is currently processing. If you need to compare a value in the hash object with one in the observation that SAS is currently processing, you must rename the data item in the hash object or rename the DATA step variable. This is important when you work with methods that modify the contents of your hash objects. Chapter 4 presents examples that modify contents of hash objects.

Examples 3.1 through 3.3 illustrate how SAS handles a DATA step variable and a hash object data item with the same name when it applies the FIND and CHECK methods.

The DATA steps in Examples 3.1 through 3.3 load data set CODELOOKUP into hash object HASHSOURCE. Data set CODELOOKUP has two observations and three variables. The second observation in CODELOOKUP has missing values for variables CODEDESC and CODEDATE. Output 3.1 lists the observations in data set CODELOOKUP.

Output 3.1 PROC PRINT of CODELOOKUP

Obs	code	codedesc	codedate
1	AB871K	Credit	02/08/2013
2	GG401P		.

All of the DATA steps in Examples 3.1 through 3.3 start with an ATTRIB statement that defines the key and data variables that are also items in hash object HASHSOURCE. Next, SAS loads data set CODELOOKUP into HASHSOURCE. None of the DATA steps in the three examples reads in a data set with a SET statement. The examples instead create observations by executing DATA steps.

Illustrating How the FIND Method Overwrites Values of DATA Step Variables

Example 3.1 creates two observations whose values for CODE are the same as those in the two entries in hash object HASHSOURCE. The values assigned to CODEDESC and CODEDATE are different than the values in HASHSOURCE. The two OUTPUT statements write the two observations that the DATA step creates to data set UPDATES1.

The DATA step calls the FIND method for each of the two values of CODE. Since both key values are present in HASHSOURCE, SAS overwrites the values that the DATA step assigned to these variables with the values it retrieves from HASHSOURCE for data items CODEDESC and CODEDATE. This includes the replacement of nonmissing values with missing values for variables CODEDESC and CODEDATE for CODE='GG401P'.

Example 3.1 Illustrating How the FIND Method Overwrites Values of DATA Step Variables

```
data updates1;
  attrib code length=$6
         codedesc length=$25
         codedate length=8 format=mmddyy10.;

  declare hash hashsource(dataset: 'codelookup',ordered: 'y');
  hashsource.definekey('code');
  hashsource.definedata('code','codedesc','codedate');
  hashsource.definedone();

  call missing(code,codedesc,codedate);

  code='AB871K';
  codedesc='Debit';
  codedate='15feb2013'd;
  rc=hashsource.find();
  output;
  code='GG401P';
  codedesc='Credit';
  codedate='23feb2013'd;
  rc=hashsource.find();
  output;
run;
```

Output 3.2 shows that data set UPDATES1 contains the values of CODEDESC and CODEDATE that were in hash object HASHSOURCE. Note that the values of CODEDESC and CODEDATE for CODE='GG401P' are missing. The missing values that SAS retrieves from HASHSOURCE for this value of CODE overwrite the non-missing values assigned by the DATA step assignment statements.

Output 3.2 PROC PRINT of UPDATES1

Obs	code	codedesc	codedate	rc
1	AB871K	Credit	02/08/2013	0
2	GG401P		.	0

Illustrating How the CHECK Method Does Not Overwrite DATA Step Variable Values

When you replace the calls to the FIND method in the DATA step in Example 3.1 with calls to the CHECK method, SAS does not overwrite the values of DATA step variables CODEDESC and CODEDATE. The CHECK method only looks for the presence of a key value; it does not return any data from the hash object. Example 3.2 lists the statements that follow the hash definition in Example 3.1 where the statements from Example 3.1 that contain the CHECK method are replaced with calls to the FIND method. Example 3.2 creates data set UPDATES2.

Example 3.2 Illustrating How the CHECK Method Does Not Overwrite DATA Step Variable Values

```
code='AB871K';
codedesc='Debit';
codedate='15feb2013'd;
rc=hashsource.check();
output;
code='GG401P';
codedesc='Credit';
codedate='23feb2013'd;
rc=hashsource.check();
output;
```

Output 3.3 lists the contents of data set UPDATES2 after executing the DATA step in Example 3.2 that makes the two calls to the CHECK method.

Output 3.3 PROC PRINT of UPDATES2

Obs	code	codedesc	codedate	rc
1	AB871K	Debit	02/15/2013	0
2	GG401P	Credit	02/23/2013	0

Illustrating the FIND Method When Data Items Are Renamed

In Example 3.3, data set options rename variables CODEDESC and CODEDATE when SAS loads data set CODELOOKUP into hash object HASHSOURCE. When SAS applies the FIND method in Example 3.3, the values for CODEDESC and CODEDATE that are in HASHSOURCE do not replace the values of DATA step variables CODEDESC and CODEDATE.

If you rename these same-named data items, you can add SAS language statements that process and test both the DATA step variable values and the hash object item values in your DATA step.

The ATTRIB statement includes the names of the renamed variables. These new variables become part of data set UPDATES3 as shown in Output 3.4.

Example 3.3 Illustrating the FIND Method When Data Items Are Renamed

```
data updates3;
  attrib code length=$6
         codedesc length=$25
         codedesc_hash length=$25
         codedate length=8 format=mmddyy10.
         codedate_hash length=8 format=mmddyy10.;

  declare hash hashsource(dataset: 'codelookup(rename=
         (codedesc=codedesc_hash codedate=codedate_hash))',
          ordered: 'y');
  hashsource.definekey('code');
  hashsource.definedata('code','codedesc_hash','codedate_hash');
  hashsource.definedone();
  call missing(code,codedesc_hash,codedate_hash);

  code='AB871K';
  codedesc='Debit';
  codedate='15feb2013'd;
  rc=hashsource.find();
  output;
  code='GG401P';
  codedesc='Credit';
  codedate='23feb2013'd;
  rc=hashsource.find();
  output;
run;
```

Output 3.4 shows the values for CODEDESC_HASH and CODEDATE_HASH retrieved from HASHSOURCE as well as the values for CODEDESC and CODEDATE that the assignment statements specify.

Output 3.4 PROC PRINT of UPDATES3

Obs	code	codedesc	codedesc_hash	codedate	codedate_hash	rc
1	AB871K	Debit	Credit	02/15/2013	02/08/2013	0
2	GG401P	Credit		02/23/2013	.	0

Application Example: Verifying Presence of Key Values

Example 3.4 demonstrates how you can use a hash object to verify presence or absence of a key value in the hash object that serves as your lookup table. The application does not require that you return associated data items.

Data set PATIENTS_TODAY contains the schedule of patients who have medical appointments on July 9, 2013. The goal is to find the patients who have not updated their medical history within the past year, or who have no history data. Data set PATIENT_HISTORY maintains this information for all patients, and this data set contains 100,000 observations. Output 3.5 shows the first 15 observations from data set PATIENT_HISTORY.

Output 3.5 PROC PRINT of PATIENT_HISTORY (first 15 observations)

Obs	ptid	ptln	ptfn	ptmi	gender	dob	pthistdate
1	HKLGBJKJ	Alexander	Brooke	D	F	05/14/1948	01/04/2011
2	G6N17AIV	Coleman	Richard	Q	F	12/13/1965	03/17/2010
3	9MH1D694	Butler	Jacqueline	I	F	07/19/1966	.
4	MSG50T7R	Miller	Jose	Z	M	09/03/1971	12/16/2012
5	3X7WU4PE	Taylor	Alexandra	W	F	02/04/1981	12/21/2009
6	JRBHYGLT	Washington	Katie	Y	M	07/20/1986	03/15/2012
7	LRYU6GRI	Murphy	Dakota	U	F	03/06/2011	06/16/2012
8	2TVEU84S	Clark	Isaac	W	F	07/14/1948	01/04/2009
9	OM5Y0OTT	Davis	Kaitlyn	S	F	11/07/1940	07/27/2012
10	WY16MIPR	Edwards	Crystal	Z	F	02/16/1926	11/19/2011
11	Z5J33NZ6	Phillips	Isaac	J	M	11/05/1985	05/25/2008
12	71LMG430	Jackson	Logan	D	F	09/10/1993	09/21/2010
13	U9Q2R1MS	Brown	Kaitlyn	J	F	11/16/1986	05/25/2011
14	NIMZF2V1	Bennett	Antonio	G	F	05/27/1984	01/15/2012
15	XR1ACQCR	Butler	Amy	X	F	12/17/1984	10/17/2010

Output 3.6 lists the eight observations in PATIENTS_TODAY. Output 3.5 shows the history data for three of these eight patients: Miller, Phillips, and Jacqueline Butler. Miller's history date is within one year of the appointment date. Phillips' last history date is more than one year ago. Jacqueline Butler has no history date.

Output 3.6 PROC PRINT of PATIENTS_TODAY

Obs	ptid	ptln	ptfn	ptmi	appt
1	MSG50T7R	Miller	Jose	Z	09JUL13:09:20
2	WRYI8G3P	Jones	Kaitlyn	R	09JUL13:09:40
3	ZD48VQOV	Hernandez	Jessica	U	09JUL13:10:00
4	932UD3PA	Smith	Kathryn	E	09JUL13:10:20
5	Z5J33NZ6	Phillips	Isaac	J	09JUL13:10:40
6	9MH1D694	Butler	Jacqueline	I	09JUL13:11:00
7	D3AQ3F79	Martinez	Justin	J	09JUL13:11:20
8	EENQO8XO	Green	Ethan	M	09JUL13:11:40

Example 3.4 starts by loading hash object HISTDATES with data from data set PATIENT_HISTORY for patients whose history information was updated within the past 365 days. Hash object HISTDATES is keyed by patient ID, and it contains no data items.

The DATA step processes PATIENTS_TODAY sequentially. The DATA step looks up each patient's ID in HISTDATES with the CHECK method. If the patient's ID is not found in HISTDATES, the conclusion is that either the patient is new or the patient does not have an updated history in the past 365 days.

SAS returns a non-zero return code if the CHECK method does not find the patient's ID in HISTDATES. SAS outputs observations to NEEDS_UPDATE if the CHECK method returns a non-zero return code. Data set NEEDS_UPDATE contains the IDs and names of the patients who require an updated medical history.

Since this DATA step does not need to return any data items from HISTORY, the CHECK method can be used. In contrast, use the FIND method if you need to return data items.

Example 3.4 Verifying Presence or Absence of a Key Value

```
data needs_update;
  if _n_=1 then do;
    declare hash histdates(dataset: "patient_history(
                                     where=(pthistdate ge
                                     '09jul2012'd)");

    histdates.definekey('ptid');
    histdates.defineDone();
  end;

  set patients_today;
  drop rc;

  rc=histdates.check();

  if rc ne 0 then output needs_update;
run;
```

Output 3.7 lists the observations in NEEDS_UPDATE. This is a list of the patients on the schedule who have not had history information recorded within the past year.

Output 3.7 PROC PRINT of NEEDS_UPDATE

Obs	ptid	ptln	ptfn	ptmi	appt
1	WRYI8G3P	Jones	Kaitlyn	R	09JUL13:09:40
2	ZD48VQOV	Hernandez	Jessica	U	09JUL13:10:00
3	932UD3PA	Smith	Kathryn	E	09JUL13:10:20
4	Z5J33NZ6	Phillips	Isaac	J	09JUL13:10:40
5	9MH1D694	Butler	Jacqueline	I	09JUL13:11:00
6	D3AQ3F79	Martinez	Justin	J	09JUL13:11:20
7	EENQO8XO	Green	Ethan	M	09JUL13:11:40

Application Example: Returning Data Items from a Hash Object

Example 3.5 looks up key values in a hash object that defines data items. When the lookup finds the key value in the hash object, the FIND method returns data from the hash object as well.

Data set EMPHOURS contains a list of the hours worked by six employees. The employees are identified by employee ID. The goal in this example is to find the name of each employee and each employee's pay level. Data set EMPLOYEES stores the employee names and pay levels and also contains employee ID.

Output 3.8 lists the six observations in EMPHOURS.

Output 3.8 PROC PRINT of EMPHOURS

Obs	empid	emphours
1	6XBIFI	38.5
2	WA4D7N	22.0
3	VPA9EF	43.0
4	TZ6OUB	11.5
5	L6KKHS	29.0
6	8TN7WL	38.0

Output 3.9 lists the first 20 observations in EMPLOYEES. Five of the six observations in EMPHOURS have a row in Output 3.9. EMPID="L6KKHS" does not.

Output 3.9 PROC PRINT of EMPLOYEES (first 20 observations)

Obs	empid	empln	empfn	empmi	gender	startdate	emppaylevel
1	6XBIFI	Ramirez	Danielle	N	F	04/21/1989	AIb
2	AWIUME	Thompson	Catherine	D	F	06/18/1986	PIIIa
3	06KH8Q	Chang	William	T	M	07/23/2002	PIIa
4	WA4D7N	Garcia	Breanna	X	F	08/20/1982	AIb
5	OOQT3Z	Jones	Brooke	E	F	08/28/1994	MIIa
6	1JU28B	Smith	Matthew	I	M	08/22/1982	TIIIb
7	V8OARE	Hall	Samuel	B	M	05/25/2010	PIb
8	1GTXQ2	Parker	Nathaniel	S	M	08/12/1996	PIc
9	VPA9EF	Baker	Cheyenne	C	F	02/24/1990	AIIa
10	0IP7L6	Hughes	Alexander	N	M	08/08/1991	TIIb
11	Q1A4SU	Sanchez	Nathaniel	W	M	08/13/1998	TIId
12	ANWFGX	Green	Tyler	I	M	12/04/1991	TIc
13	L1I8Y7	Edwards	Angelica	O	F	11/18/1991	MIIIa
14	TZ6OUB	White	Heather	T	F	01/27/1999	AIIIa
15	235TWE	King	Briana	M	F	10/08/1992	TIIc
16	XYOJC7	Scott	Mark	T	M	01/15/2002	TIIa
17	8TN7WL	Miller	Tyler	J	M	08/31/1998	AIIIc
18	US3DZP	Brown	Sarah	U	F	12/11/2000	TIb
19	ODBAIZ	Jones	Rachel	T	F	10/11/1999	PIIb
20	A4GJG4	Johnson	Angelica	Z	F	01/01/1994	AId

The DATA step creates hash object EMP and loads data into EMP from variables EMPLN, EMPFN, EMPMI, and EMPPAYLEVEL in data set EMPLOYEES. It defines variable EMPID as the key item in hash object EMP.

The DATA step processes data set EMPHOURS sequentially. It looks up information about each employee in hash object EMP. Output data set EMPINFO contains the variables from EMPHOURS plus the variables corresponding to the data items retrieved from hash object EMP.

The ATTRIB statement at the beginning of the DATA step defines the variables that serve as the data items. If these variables were not defined, the CALL MISSING statement would cause SAS to define them using the default of an 8-byte numeric variable. The three variables that define the three parts of the name must contain character-type data. Therefore, an ATTRIB statement or LENGTH statement must precede the CALL MISSING statement. Without one of these two statements, Example 3.5 would end in error.

The DATA step does not do any special processing if the EMPID value does not exist in hash object EMP. In your application, you may want to add code that tests the return code from the

FIND method and execute statements based on the success of the FIND method. Example 3.5 saves the return code value in variable RC, which you can see in Output 3.10. The value of RC is 0 for the five observations in EMPHOURS where the value of EMPID is present in EMP. The RC value is not zero for EMPID='L6KKHS', which is not present in EMP.

Example 3.6 includes code that branches the processing based on the value of the return code from the FIND method.

Example 3.5 Returning Data Items from a Hash Object

```
data empinfo;
  attrib empln length=$30 label='Employee Last Name'
         empfn length=$25 label='Employee First Name'
         empmi length=$1  label='Employee Middle Initial'
         emppaylevel length=$4  label='Pay level';

  if _n_=1 then do;
    declare hash emp(dataset: "employees");
    emp.definekey('empid');
    emp.definedata('empln','empfn','empmi','emppaylevel');
    emp.definedone();
    call missing(empln, empfn, empmi, emppaylevel);
  end;

  set emphours;

  rc=emp.find();
run;
```

Output 3.10 lists the observations in EMPINFO.

Output 3.10 PROC PRINT of EMPINFO

Obs	empln	empfn	empmi	emppaylevel	empid	emphours	rc
1	Ramirez	Danielle	N	AIb	6XBIFI	38.5	0
2	Garcia	Breanna	X	AIb	WA4D7N	22.0	0
3	Baker	Cheyenne	C	AIIa	VPA9EF	43.0	0
4	White	Heather	T	AIII	TZ6OUB	11.5	0
5					L6KKHS	29.0	160038
6	Miller	Tyler	J	AIII	8TN7WL	38.0	0

Application Example: Defining the Key Value During Processing of the DATA Step

Example 3.6 shows that it is possible to define the key variable value with SAS language statements in the same DATA step that looks up the key value in a hash object. The previous examples in this chapter use key values that either are stored in the data set that the DATA step read sequentially or are specified as literal values.

The goal of Example 3.6 is to find the prize to award to the persons in PRIZEWINNERS. The prizes are categorized by age and type. The two age groups are adults and children. The three types of prizes are fitness classes ('class'), sport facilities passes ('pass'), and sports equipment ('sports'). Data set PRIZECATALOG stores the available prizes.

Output 3.11 lists the observations in PRIZEWINNERS.

Output 3.11 PROC PRINT of PRIZEWINNERS

Obs	winnername	age	prizetype	prizelevel
1	Allison Johnson	32	class	3
2	Jeffrey Davis	28	pass	2
3	Brady Hall	10	pass	3
4	Courtney Hill	16	sports	2
5	Tiffany Scott	41	class	4
6	Teresa Martinez	18	pass	4
7	Patricia Rodriguez	48	class	1
8	Samantha Washington	12	class	2

Output 3.12 lists the observations in PRIZECATALOG.

Output 3.12 PROC PRINT of PRIZECATALOG

Obs	prizecode	prize
1	ASPORTS1	Heart Rate Monitor
2	ASPORTS2	Walking Poles
3	ASPORTS3	Free Weights
4	CSPORTS1	Kick Scooter
5	CSPORTS2	Inline Skates
6	CSPORTS3	Skateboard
7	APASS1	Athletic Center Pass
8	APASS2	Hiking Trails Pass
9	APASS3	Swim Center Pass
10	CPASS1	Indoor Playground Pass

Obs	prizecode	prize
11	CPASS2	Swim Center Pass
12	CPASS3	Skateboard Park Pass
13	CLASS1	Golf Lessons
14	CLASS2	Tennis Lessons
15	CLASS3	Fitness Class
16	CLASS4	Yoga Class

Example 3.6 starts by loading data set PRIZECATALOG into hash object P. Its key item is PRIZECODE. Its one data item is the prize description stored in variable PRIZE.

This DATA step reads data set PRIZEWINNERS sequentially. Variable PRIZECODE does not exist in data set PRIZEWINNERS, but the values of PRIZECODE are the key values in hash object P. The DATA step determines the prize code for each observation in PRIZEWINNERS by concatenating two or three variables.

Most prizes are awarded based on the age of the person. Codes for prizes awarded to adults start with “A”. Codes for prizes awarded to children start with “C”. Prizes not determined by age do not have a prefix letter. The four fitness class prizes fall into this last category. The rest of the value of PRIZECODE is the result of concatenating the values of the two variables PRIZETYPE and PRIZELEVEL.

IF-THEN statements define the value of PRIZECODE for each observation in PRIZEWINNERS. When SAS does not find a PRIZECODE value in hash object P, as is true for the observation for Teresa Martinez, Example 3.6 assigns default text to variable PRIZE.

Example 3.6 Defining the Key Value During Processing of the DATA Step

```
data prizelist;
  keep winnername age prize;
  attrib winnername length=$20 label='Name of Winner'
         prizecode length=$8 label='Prize Code'
         prize length=$40 label='Prize Description';

  if _n_=1 then do;
    declare hash p(dataset:'prizecatalog');
    p.defineKey('prizecode');
    p.defineData('prize');
    p.defineDone();
    call missing(prize);
  end;

  set prizewinners;
```

```
    if upcase(prizetype) ne 'CLASS' then do;
      if age ge 18 then
            prizecode=cats('A',upcase(prizetype),put(prizelevel,1.));
      else if age ne . then
            prizecode=cats('C',upcase(prizetype),put(prizelevel,1.));
    end;
    else prizecode=cats(upcase(prizetype),put(prizelevel,1.));

    rc=p.find();
    if rc ne 0 then prize=catx(' ',"*** Unknown Prize Code:",prizecode);
  run;
```

Output 3.13 lists the observations in PRIZELIST.

Output 3.13 PROC PRINT of PRIZELIST

Obs	winnername	prize	age
1	Allison Johnson	Fitness Class	32
2	Jeffrey Davis	Hiking Trails Pass	28
3	Brady Hall	Skateboard Park Pass	10
4	Courtney Hill	Inline Skates	16
5	Tiffany Scott	Yoga Class	41
6	Teresa Martinez	*** Unknown Prize Code: APASS4	18
7	Patricia Rodriguez	Golf Lessons	48
8	Samantha Washington	Tennis Lessons	12

Application Example: Searching for a Key Value in Multiple Hash Objects

Example 3.7 shows how to define more than one hash object in a DATA step and search each of the hash objects using the same key value. Two reasons you might need multiple hash objects open in the same DATA step are:

- You need to find the presence or absence of a key in several lookup tables. Example 3.7 demonstrates this application.
- You have multiple lookup tables and you are not sure in which source a key will be found.

The DATA step in Example 3.7 has two goals:

1. Determine which physicians in data set DOCTORLIST work at two or more locations.
2. Determine which physicians do not work at any of the three locations.

The data sets DOCTORS_SOUTHSIDE, DOCTORS_MAPLEWOOD, and DOCTORS_MIDTOWN contain employee IDs for physicians who work at each of the three locations, Southside, Maplewood, and Midtown. Output 3.14 lists the contents of these three data sets.

Output 3.14 PROC PRINTs of DOCTORS_SOUTHSIDE, DOCTORS_MAPLEWOOD, and DOCTORS_MIDTOWN

SOUTHSIDE	
Obs	empid
1	3D401A
2	NBJ588
3	WE3HJH

MAPLEWOOD	
Obs	empid
1	3D401A
2	LDT47L
3	NBK588
4	O0OBRU

MIDTOWN	
Obs	empid
1	2VIHPJ
2	3D401A
3	HPM27C
4	NBK588
5	NTWJLF
6	S7JHOM

Output 3.15 lists the observations in data set DOCTORLIST. EMPID values BKK94F and MN2ZY6 are not present in any of the three data sets shown in Output 3.14.

Output 3.15 PROC PRINT of DOCTORLIST

Obs	empid	empln	empfn	empmi
1	2VIHPJ	Thomas	Noah	S
2	3D401A	Long	Alicia	E
3	BKK94F	Robinson	Michael	M
4	HPM27C	Cooper	Brian	F
5	LDT47L	Torres	Chelsea	Q
6	MN2ZY6	Anderson	Chase	P
7	NBK588	Scott	Kyle	S
8	O0OBRU	Cooper	Bradley	D
9	WE3HJH	Peterson	Jamie	Z

The DATA step starts by creating three hash objects, SS, MP, and MT, and loads data into each of them from each of the three location data sets. For each observation in DOCTORLIST, SAS uses the CHECK method to look for the current value of EMPID in each of the three hash objects. The CHECK method is used because it is not necessary to return any data from any of the three hash objects. The only goal is to determine the number of times the value of EMPID is found in the three hash objects.

The code outputs an observation to data set MULTWORKLOCS if the current value of EMPID is found in more than one of the three hash objects. The code outputs an observation to data set NOWORKLOC if the current value of EMPID is not found in any of the three hash objects.

For observations for physicians working at multiple locations, new character variable LOCATIONS contains the concatenated list of location names. SAS assigns default text to variable LOCATIONS for observations for physicians who do not work at any of the three locations.

Variable NLOCS contains the number of locations at which the physician works. Its value is determined by evaluating the return codes, SSRC, MWRC, and MTRC, from each of the three lookups in the three hash tables. A return code of 0 means the key value was found in the hash object.

Example 3.7 Searching for a Key Value in Multiple Hash Objects

```
data multworklocs noworkloc;
  if _n_=1 then do;
    declare hash ss(dataset: "doctors_southside");
    ss.definekey('empid');
    ss.definedone();

    declare hash mw(dataset: "doctors_maplewood");
    mw.definekey('empid');
    mw.definedone();

    declare hash mt(dataset: "doctors_midtown");
    mt.definekey('empid');
    mt.definedone();
  end;

  drop ssrc mwrc mtrc nlocs;

  set doctorlist;

  attrib location_list length=$50 label='Employee Work Locations';

  ssrc=ss.check();
  mwrc=mw.check();
  mtrc=mt.check();

  nlocs=(ssrc=0) + (mwrc=0) + (mtrc=0);

  if nlocs ge 2 then do;
    if ssrc=0 then location_list="Southside";
    if mwrc=0 then location_list=catx(',',location_list,"Maplewood");
    if mtrc=0 then location_list=catx(',',location_list,"Midtown");
    output multworklocs;
  end;
```

```
      else if nlocs=0 then do;
        location_list="**Not at Southside, Maplewood, or Midtown";
        output noworkloc;
      end;

    run;
```

Output 3.16 lists the contents of data set MULTWORKLOCS.

Output 3.16 PROC PRINT of MULTWORKLOCS

Obs	empid	empln	empfn	empmi	location_list
1	3D401A	Long	Alicia	E	Southside,Maplewood,Midtown
2	NBK588	Scott	Kyle	S	Maplewood,Midtown

Output 3.17 lists the contents of data set NOWORKLOC.

Output 3.17 PROC PRINT of NOWORKLOC

Obs	empid	empln	empfn	empmi	location_list
1	BKK94F	Robinson	Michael	M	**Not at Southside, Maplewood, or Midtown
2	MN2ZY6	Anderson	Chase	P	**Not at Southside, Maplewood, or Midtown

Application Example: Combining Data from Multiple Sources

When you want to match-merge several data sets and the data sets do not all match by the same variables, it can take several DATA steps and PROC SORT steps to produce the data set that you need. If you use PROC SQL, it can take several joins, and multiple joins can be complicated to write. By evaluating your data sources and the output data set that you need, it might be possible to write one DATA step that uses hash objects to combine data from multiple sources.

Example 3.8 shows that you can search multiple hash objects in a DATA step for specific key values. Further, it is not necessary for all hash objects you define in a DATA step to have the same key items. You can also load multiple hash objects in a DATA step with data from the same data set, and you can define these hash objects with different key items.

A way to approach combining data from multiple sources using hash objects is to start by determining what an observation should look like in your output data set. The easiest next step is to read from the data set that has observations in the same structure that you need in your output data set. You load the remaining data sources into hash objects. The DATA step then contains several statements that access the contents of the multiple hash objects.

Example 3.8 was written using the guidelines in the preceding paragraph. The goal of Example 3.8 is to find information about subjects at the beginning of a study, and this information is stored in several data sets.

Table 3.2 describes the four data sets that Example 3.8 accesses.

Table 3.2 Input Data Sets for Example 3.8

Data Set Name	Description	Variables Needed
STUDYSUBJECTS	One observation per subject. Variable STUDYID uniquely identifies an observation. Variable NURSEID uniquely identifies the nurse associated with the subject.	All variables
STUDYAPPTS	One observation per clinic appointment per subject. Variables STUDYID and APPTTYPE uniquely identify an observation.	All variables
STUDYSTAFF	One observation per staff member. Variable STAFFID uniquely identifies an observation.	STAFFID, STAFFNAME, SITEID
STUDYSITES	One observation per study site. Variable SITEID uniquely identifies an observation.	SITEID, SITENAME

Looking at the list of input data sets, data set STUDYSUBJECTS has the structure of one observation per subject that is required for the output data set. The data sets are related as follows:

- Data set STUDYSUBJECTS has a one-to-many relationship by STUDYID with data set STUDYAPPTS since STUDYAPPTS contains one observation per subject per appointment. The task of Example 3.8 is to select only the observations where APPTTYPE='BASELINE' from STUDYAPPTS. All subjects have only one baseline observation.
- Variable NURSEID in data set STUDYSUBJECTS links to STAFFID in data set STUDYSTAFF.
- Variable SITEID in STUDYSTAFF links to SITEID in SITES.

Output 3.18 lists the first five observations in STUDYSUBJECTS.

Output 3.18 PROC PRINT of STUDYSUBJECTS (first 5 observations)

Obs	studyid	nurseid	dieticianid	treatmentgroup
1	GHY101	461	582	A
2	REA102	391	592	B
3	PLK103	393	387	A
4	MIJ104	461	592	B
5	NHC105	240	439	A

Output 3.19 lists the observations in STUDYAPPTS for four of the five subjects listed in Output 3.18. The fifth subject, PLK103, does not have any observations in STUDYAPPTS, and this observation's value for NURSEID is not present in STAFFID.

Output 3.19 PROC PRINT of STUDYAPPTS (for subjects listed in Output 3.18)

Obs	studyid	apptdate	appttype	weight	systol	diast
35	GHY101	03/08/2013	Baseline	180	180	90
36	GHY101	04/15/2013	One Month	184	181	89
37	GHY101	07/02/2013	Two Month	186	184	90
38	GHY101	11/26/2013	Four Month	180	182	88
39	GHY101	06/05/2014	Six Month	177	170	84
50	MIJ104	02/08/2013	Baseline	205	135	90
51	MIJ104	03/22/2013	One Month	212	132	90
52	MIJ104	06/10/2013	Two Month	218	134	85
53	MIJ104	10/28/2013	Four Month	220	135	84
56	NHC105	03/15/2013	Baseline	222	148	82
57	NHC105	04/15/2013	One Month	232	154	85
58	NHC105	08/22/2013	Four Month	218	154	84
59	REA102	03/07/2013	Baseline	154	173	85
60	REA102	04/16/2013	One Month	154	178	89
61	REA102	06/27/2013	Two Month	148	170	85
62	REA102	11/08/2013	Four Month	142	159	83

Output 3.20 shows the observations in STUDYSTAFF.

Output 3.20 PROC PRINT of STUDYSTAFF

Obs	staffid	unit	staffname	degree	siteid
1	371	Nursing	EN Cameron	RN	1
2	461	Nursing	YU Garcia	CNP	4
3	104	Dietary	RD Hong	RD	1
4	240	Nursing	TR Howe	RN	3
5	439	Dietary	KA Izzo	RD	2
6	592	Dietary	HS James	RD	2
7	387	Dietary	HN Lee	RD	3
8	391	Nursing	MR Smith	CNP	2

Output 3.21 lists the observations in STUDYSITES.

Output 3.21 PROC PRINT of STUDYSITES

Obs	siteid	sitename	sitemanager
1	1	Midtown Primary Care Clinic	FD Weston
2	2	Southside Primary Care Clinic	AW Liu
3	3	Maplewood Wellness Clinic	PR Ross

Since the structure of the output data set is the same as that of STUDYSUBJECTS, the DATA step reads each observation in STUDYSUBJECTS. It defines three hash objects to contain data from the three other data sources:

- Hash object B contains data from STUDYAPPTS where APPTTYPE='BASELINE'.
- Hash object F contains data from STUDYSTAFF.
- Hash object T contains data from STUDYSITES.

On each iteration of the DATA step, SAS uses the FIND method to look for the current observation's STUDYID value in key item STUDYID in hash object B. If the value is present, the data items associated with that key value in B are returned to the same-named DATA step variables. If the key value is not present in B, which is the case for STUDYID='PLK103', SAS assigns default text to variable APPTTYPE.

Next the DATA step looks for the nurse's name and site ID in hash object F. The name of the key item in F is STAFFID. Variable NURSEID in STUDYSUBJECTS is specified on the KEY argument tag in the FIND method call to F. Since the requirement is to obtain data for only the nurse, the RENAME data set option on the DECLARE statement for F renames STAFFNAME to NURSENAME.

If SAS does not find a value for NURSEID in key item STAFFID in F, it assigns default text to NURSENAME and to SITENAME as well. Without a SITEID returned from F in this situation, the lookup in hash object T cannot occur. This is the situation for STUDYID='PLK103'.

If an entry is found in F, SAS returns the value of data item SITEID to the same-named DATA step variable. The SITEID value is now used in the last FIND method call. This last call retrieves the site name from hash object S. If SAS does not find a value for SITEID in S, it assigns default text to variable SITENAME.

Example 3.8 Combining Multiple Data Sources Using a Hash Object

```
data study_baseline;
  attrib studyid   length=$6
         appttype  length=$15
         apptdate  length=8 format=mmddyy10.
         weight    length=8
         systol    length=8
         diast     length=8
         staffid   length=8
         nursename length=$12
```

```
          siteid length=8
          sitename length=$30;

  if _n_=1 then do;
    declare hash b(dataset: "studyappts(where=(appttype='Baseline'))");
    b.defineKey('studyid');
    b.definedata(all: 'yes');
    b.definedone();

    declare hash f(dataset:"studystaff(rename=(staffname=nursename))");
    f.defineKey('staffid');
    f.definedata('siteid','nursename');
    f.definedone();

    declare hash t(dataset: "studysites");
    t.definekey('siteid');
    t.definedata('sitename');
    t.definedone();

    call missing(appttype,apptdate,weight,systol,diast,
                 staffid,nursename,siteid,sitename);
  end;

  set studysubjects;

  drop rc staffid;

  rc=b.find();
  if rc ne 0 then appttype='***No appts';
  rc=f.find(key: nurseid);
  if rc=0 then do;
    rc=t.find();
    if rc ne 0 then sitename='**Unknown';
  end;
  else if rc ne 0 then do;
    nursename='**Unknown';
    sitename='**Unknown';
  end;
run;
```

Output 3.22 lists the selected baseline data for the first five observations saved in output data set STUDY_BASELINE.

Output 3.22 PROC PRINT of STUDY_BASELINE (first 5 observations)

Obs	studyid	appttype	apptdate	weight	systol	diast	nursename	siteid	sitename	nurse id	dietician id	treat-ment group
1	GHY101	Baseline	03/08/2013	180	180	90	YU Garcia	4	**Unknown	461	582	A
2	REA102	Baseline	03/07/2013	154	173	85	MR Smith	2	Southside Primary Care Clinic	391	592	B
3	PLK103	***No appts	.	.	.	.	**Unknown	.	**Unknown	393	387	A
4	MIJ104	Baseline	02/08/2013	205	135	90	YU Garcia	4	**Unknown	461	592	B
5	NHC105	Baseline	03/15/2013	222	148	82	TR Howe	3	Maplewood Wellness Clinic	240	439	A

If you used match-merging to create data set STUDY_BASELINE, you would have to execute several PROC SORT steps and DATA steps. You would sort data sets STUDYAPPTS, STUDYSTAFF, and STUDYSITES once. You would sort and read STUDYSUBJECTS three times. The first pass of STUDYSUBJECTS would be a match-merge of it with STUDYAPPTS. After sorting the result of the first pass by NURSEID, the second match-merge DATA step would merge the data set produced by the first pass with STUDYSTAFF. Last, the data set produced by the second pass would have to be sorted by SITEID and then merged to STUDYSITES.

Using Multiple Key Items to Look Up Data

You can load a lookup data set that has more than one key item into a hash object and define multiple key items for that hash object. The key items can be any mix of character items and numeric items. The process to define and use a hash object that has multiple key items is the same as the process that you follow when you define and use a hash object that has only one key item.

For example, suppose you want to find out if a group of employees are scheduled to work specific dates and shifts at a health screening event. A data set contains the schedule information for all employees scheduled to work at the event. A second data set contains a list of employees and the date and shift for which you want to check if an employee is scheduled to work. Example 3.9 shows a DATA step that performs this multiple key item lookup.

You can load the data set with the schedules for all employees into a hash object and key this hash object by the three variables—event date, shift, and employee ID—that define the criteria for determining an employee's schedule.

The advantages described at the start of the section on using single-keyed hash objects instead of arrays or user-defined formats hold true when using multiple key items. When you use a hash object, you do not have to specify the exact number of items you will store in your hash object and you can retrieve one or more data items from the hash object.

Using a hash object when you have multiple keys can be easier than working with arrays or user-defined formats. A big advantage of hash objects is that the key items in the hash object can be any mix of character items and numeric items. With arrays, you can use only numeric index values to find elements in your array. Further, unless you have a lot of experience with multi-dimensional arrays, it can be difficult to set them up and reference them easily.

A user-defined format can work with character or numeric values. However, a user-defined format can only have one value for which you can look up one value label. (You can return multiple value labels per value when you use PROC TABULATE and PROC REPORT if you have defined the format with the MULTILABEL option.) To handle multiple key items, you may have to define several formats and add programming statements that process the value label information that your formats return.

Example 3.9 checks whether a group of employees are scheduled to work specific dates and shifts at a health screening event on October 22 and 23, 2013. Data set FINDSCHED stores information to look up for a group of employees whose schedules are to be examined. Output 3.23 lists the observations in FINDSCHED.

Output 3.23 PROC PRINT of FINDSCHED

Obs	eventdate	shift	empid	empname
1	10/22/2013	AM	LISUUC	Davis, Laura W.
2	10/22/2013	AM	OYMEE3	Thomas, Hannah I.
3	10/22/2013	PM	OYMEE3	Thomas, Hannah I.
4	10/22/2013	PM	8I4YKY	Brooks, Shelby B.
5	10/23/2013	AM	7YN5NV	Perez, Paige O.
6	10/22/2013	PM	8VLUT7	Hall, Kelsey H.
7	10/23/2013	PM	8VLUT7	Hall, Kelsey H.

Data set OCTOBEREVENT contains the schedule information for the 52 employees scheduled to work at the health screening event. Employees are assigned to a morning ("AM") or afternoon ("PM") shift to perform one of six tasks: cholesterol screening; counseling; dietary advice; immunization; pharmaceutical counseling; and measurement of vital health indicators such as blood pressure and weight. Output 3.24 shows the first 30 observations in OCTOBEREVENT.

Output 3.24 PROC PRINT of OCTOBEREVENT (first 30 observations)

Obs	eventdate	shift	activity	empid	empname
1	10/22/2013	AM	Cholesterol	02QYJG	Howard, Rachel Y.
2	10/22/2013	AM	Cholesterol	8I4YKY	Brooks, Shelby B.
3	10/22/2013	AM	Cholesterol	ADHW3A	Moore, Allison W.
4	10/22/2013	AM	Dietary	IFQZ8S	Nelson, Kelly U.
5	10/22/2013	AM	Dietary	OYMEE3	Thomas, Hannah I.
6	10/22/2013	AM	Dietary	PTBHUP	Thompson, Olivia P.
7	10/22/2013	AM	Pharmaceutical	WVV7PT	Patterson, Vanessa N.
8	10/22/2013	AM	Counseling	14ZN75	Miller, Sierra Q.
9	10/22/2013	AM	Counseling	5KA7JH	Martin, Aaron G.
10	10/22/2013	AM	Counseling	BA8CRZ	Baker, Marissa R.
11	10/22/2013	AM	Counseling	FAL5UZ	Butler, Haley P.
12	10/22/2013	AM	Counseling	TMJTVP	Edwards, Brittany O.
13	10/22/2013	AM	Counseling	YXW78P	Davis, Andrea Y.
14	10/22/2013	AM	Immunization	1CV01H	Brown, Lindsey V.
15	10/22/2013	AM	Immunization	8EHCSX	Hernandez, Elizabeth S.
16	10/22/2013	AM	Immunization	HD8ERT	Paine, Mike J.
17	10/22/2013	AM	Immunization	KRTV2T	Gonzalez, Alejandro F.
18	10/22/2013	AM	Vitals	NCPO4E	Richardson, Bailey U.
19	10/22/2013	AM	Pharmaceutical	P438U8	Price, Katelyn Q.
20	10/22/2013	AM	Vitals	2S57ZI	Thompson, Alexandria B.
21	10/22/2013	AM	Vitals	69BWGZ	Jones, Marissa W.
22	10/22/2013	AM	Vitals	F97SKU	Lee, Alicia Q.
23	10/22/2013	AM	Vitals	FTZMFI	Rodriguez, Samantha E.
24	10/22/2013	AM	Vitals	OV9K5X	Baker, Jasmine O.
25	10/22/2013	PM	Cholesterol	02QYJG	Howard, Rachel Y.
26	10/22/2013	PM	Cholesterol	2S57ZI	Thompson, Alexandria B.
27	10/22/2013	PM	Cholesterol	8I4YKY	Brooks, Shelby B.
28	10/22/2013	PM	Dietary	C800UH	Patterson, Jamie Q.
29	10/22/2013	PM	Dietary	GWS2QS	Parker, Brittany D.
30	10/22/2013	PM	Dietary	LISUUC	Davis, Laura W.

The goal of Example 3.9 is to look for a match in OCTOBEREVENT for each employee and his or her schedule in FINDSCHED. The lookup is based on the values of three variables: EVENTDATE, SHIFT, and EMPID.

SAS loads data set OCTOBEREVENT into hash object S and defines the key items as the three variables that define the lookup. Hash object S has one data item, ACTIVITY. If SAS finds all three key values for an observation in FINDSCHED in hash object S, it returns the employee's task assignment to variable ACTIVITY. If SAS does not find a match in S for an observation it reads from FINDSCHED, the assumption is that the employee is not scheduled for that specific date and shift and the code assigns the default text "**Not scheduled" to variable ACTIVITY.

Example 3.9 Using Multiple Key Items to Look Up Data

```
data selectedemps;
  attrib eventdate format=mmddyy10.
         shift length=$2 label='Event Shift'
         activity length=$20 label='Event Activity'
         empid length=$6 label='Employee ID';

  if _n_=1 then do;
    declare hash s(dataset: 'octoberevent');
    s.definekey('eventdate','shift','empid');
    s.definedata('activity');
    s.definedone();

    call missing(activity);
  end;

  set findsched;
  drop rc;
  rc=s.find();
  if rc ne 0 then activity="** Not scheduled";;
run;
```

The DATA step outputs all observations in FINDSCHED to data set SELECTEDEMPS. Output 3.25 shows the contents of data set SELECTEDEMPS.

Output 3.25 PROC PRINT of SELECTEDEMPS

Obs	eventdate	shift	activity	empid	empname
1	10/22/2013	AM	** Not scheduled	LISUUC	Davis, Laura W.
2	10/22/2013	AM	Dietary	OYMEE3	Thomas, Hannah I.
3	10/22/2013	PM	Counseling	OYMEE3	Thomas, Hannah I.
4	10/22/2013	PM	Cholesterol	8I4YKY	Brooks, Shelby B.
5	10/23/2013	AM	** Not scheduled	7YN5NV	Perez, Paige O.
6	10/22/2013	PM	Immunization	8VLUT7	Hall, Kelsey H.
7	10/23/2013	PM	** Not scheduled	8VLUT7	Hall, Kelsey H.

Traversing Hash Objects

The previous examples looked for a single entry in a hash object based on key values. Sometimes you may need to sequentially read several entries in your hash object starting from a specific location or key value in your hash object. When you want to traverse a hash object in either forward or backward direction, you can define a hash iterator object to manage this iterative process.

The hash iterator object is a DATA step component object. It does not store data like a hash object. Instead it provides a way to access data in a hash object. Several methods exist that control movement through a hash object when you access it with a hash iterator object. You can start at the beginning or at the end of your hash object, or you can start the search at a specific key value. You can move through the hash object one entry at a time in no specific order, or you can request that SAS return data in order by the values of the key item.

A hash iterator object may enable you to combine in one DATA step the several steps that you would otherwise have to do. For example, with a hash iterator object applied to a hash object that has been defined with the ORDERED argument tag, you can process the data in the hash object in sorted order in a sequential way that looks like BY-group processing. Without using a hash iterator object, if you needed to do BY-group processing, you would sort your data set with PROC SORT or index your data set by the BY-variables before your DATA step processes the ordered observations. Or, you might write a PROC SQL SELECT clause that orders its selections when it outputs a table. You may need to follow your PROC SQL step with a DATA step that has SAS language statements that complete the processing of your data.

Specifying a Hash Iterator Object

When you want to traverse a hash object with a hash iterator object, the first step is to define your hash object and identify key items and data items in the same way as previous examples have shown. If you want SAS to return data in order by the values of the keys, you must add the ORDERED: argument tag to the DECLARE statement and specify its values as "YES", "ASCENDING", or "DESCENDING". Your code can still traverse a hash object if you do not specify the ORDERED: argument tag. However, in this situation, SAS does not retrieve key values from your hash object in either ascending or descending order.

The second action to take in creating a hash iterator object is to add a second DECLARE statement that defines the hash iterator object and associates it with your hash object. This is the only definition-type statement you need to create a hash iterator object. The only argument you supply on this second DECLARE statement is the name of the hash object that you want to traverse. You must always specify the name of the hash object that you want SAS to associate with the hash iterator object, and you must enclose that name in quotation marks.

The syntax of the DECLARE statement that defines a hash iterator object follows. The object-type keyword HITER always immediately follows the DECLARE keyword.

```
DECLARE HITER iterator-object-name('hash-object-name');
```

Understanding the Methods That Control Traversal of a Hash Object

Five methods—FIRST, LAST, NEXT, PREV, and SETCUR—control how the hash iterator object traverses the hash object. These methods allow you to start at the beginning or end of a hash object and move either forward or backward through the hash object. Traversal of the hash object can also start in a specific location based on a key value. From that known position in the hash object, traversal can be in either the forward or backward direction.

The syntax for these five methods follows:

```
rc=iterator-object-name.FIRST();

rc=iterator-object-name.LAST();

rc=iterator-object-name.NEXT();

rc=iterator-object-name.PREV();

rc=iterator-object-name.SETCUR(KEY: 'keyvalue-1'<,...,
            KEY: 'keyvalue-n'>);
```

The FIRST, LAST, NEXT, and PREV methods have no argument tags. The FIRST method returns the first key value in the hash object, and the LAST method returns the last key value. The NEXT method moves forward in the hash object and returns the next key value in the hash object. The PREV method moves backward in the hash object and returns the previous key value.

You can traverse a hash object with these methods even if you do not define it as an ordered hash object. In this situation, SAS does not return key values in ascending order, descending order, or in the order SAS added the key values to the hash object.

The KEY argument tag on the SETCUR method allows you to specify a starting key value from which to begin the traversal of the underlying hash object. For example, suppose your hash object is keyed by date and your task is to find data in the hash object from a specific date value forward. If you define your hash object with the ORDERED: “YES” argument tag, you could use the SETCUR method to start the search at that specific date and move forward in time through the hash object from that starting date.

A limitation of the SETCUR method is that the key value you specify must exist in the hash object that your hash iterator object traverses. For example, if you specify a date on the SETCUR KEY argument tag that is not in the hash object, SAS does not move to the nearest date. It sets a non-zero return code as the result of the SETCUR method call.

The SETCUR method syntax shows quotation marks around the key-value tags. Use the quotation marks only when you specify a literal character value.

Hash objects can have multiple sets of data items per key value. When using a hash iterator object in that situation, you use a different set of methods than the ones listed above. Chapter 5 presents examples that work with hash objects that can have multiple sets of data items per key value.

Illustrating How the Hash Iterator Object Traverses a Hash Object

This section presents two examples that illustrate how a hash iterator object traverses a hash object and retrieves data from the hash object. Examples 3.10 and 3.11 process data set BPSTUDY, which has observations for two patients whose weight and blood pressure were measured on the first Wednesday of every month in 2013. Output 3.26 shows the contents of BPSTUDY.

Output 3.26 PROC PRINT of BPSTUDY

Obs	ptid	examdate	weight	systolic	diastolic
1	V8810YQ4	01/02/2013	182	154	89
2	V8810YQ4	02/06/2013	175	148	83
3	V8810YQ4	03/06/2013	170	147	79
4	V8810YQ4	04/03/2013	158	135	78
5	V8810YQ4	05/01/2013	155	133	79
6	V8810YQ4	06/05/2013	154	129	75
7	V8810YQ4	07/03/2013	148	131	80
8	V8810YQ4	08/07/2013	150	128	75
9	V8810YQ4	09/04/2013	149	129	72
10	V8810YQ4	10/02/2013	146	130	77
11	V8810YQ4	11/06/2013	140	126	78
12	V8810YQ4	12/04/2013	146	122	80
13	56KJR3D9	01/02/2013	142	147	82
14	56KJR3D9	02/06/2013	144	138	90
15	56KJR3D9	03/06/2013	139	133	77
16	56KJR3D9	04/03/2013	133	141	79
17	56KJR3D9	05/01/2013	131	124	74
18	56KJR3D9	06/05/2013	134	120	80

Obs	ptid	examdate	weight	systolic	diastolic
19	56KJR3D9	07/03/2013	133	124	81
20	56KJR3D9	08/07/2013	130	125	74
21	56KJR3D9	09/04/2013	129	123	71
22	56KJR3D9	10/02/2013	129	119	72
23	56KJR3D9	11/06/2013	126	120	72
24	56KJR3D9	12/04/2013	124	118	70

Using the ORDERED Argument Tag When a Hash Iterator Object Traverses a Hash Object

When you define a hash object, you must always specify at least one key item. When you include the ORDERED: argument tag on the associated DECLARE statement (and do not specify ORDERED: "NO"), SAS retrieves data from the hash object in order by the values of the key items. Without this specification, and even if the data set you load into the hash object is in order by the key values, SAS does not necessarily retrieve data from the hash object in key value order.

Example 3.10 loads data set BPSTUDY into hash object BPCK, which has two key items: PTID and EXAMDATE. The goal is to find dates when the patient's systolic blood pressure is greater than 140. The ORDERED: argument tag is not included on the DECLARE statement for BPCK. The DATA step defines hash iterator object BPCKITER and associates it with BPCK.

The DATA step iterates only once. The statement with the FIRST method call to BPCKITER returns the first entry in BPCK. The NEXT method call to BPCKITER in the DO UNTIL block reads all of the remaining entries in hash object BPCK. SAS outputs only the observations where the value of SYSTOLIC is greater than 140.

Example 3.10 Using a Hash Iterator Object and Retrieving Data from a Hash Object When the ORDERED: Argument Tag Is Not Used

```
data highsystolic;
  attrib ptid length=$8
         examdate format=mmddyy10.
         weight length=8
         systolic length=8
         diastolic length=8;

  declare hash bpck(dataset: 'bpstudy');
  declare hiter bpckiter('bpck');
  bpck.definekey('ptid','examdate');
  bpck.definedata('ptid','examdate','weight','systolic','diastolic');
  bpck.definedone();

  call missing(ptid,examdate,weight,systolic,diastolic);

  drop rc;
```

```
    rc=bpckiter.first();
    do until (rc ne 0);
      if systolic gt 140 then output;
      rc=bpckiter.next();
    end;
  run;
```

Output 3.27 shows that while SAS finds all systolic pressures greater than 140, it does not retrieve data from BPCK in order by the values of key items PTID and EXAMDATE.

Output 3.27 PROC PRINT of HIGHSYSTOLIC

Obs	ptid	examdate	weight	systolic	diastolic
1	V8810YQ4	02/06/2013	175	148	83
2	V8810YQ4	01/02/2013	182	154	89
3	56KJR3D9	01/02/2013	142	147	82
4	V8810YQ4	03/06/2013	170	147	79
5	56KJR3D9	04/03/2013	133	141	79

By adding the ORDERED: 'YES' argument tag to the first DECLARE statement, SAS now retrieves items from BPCK in order by the values of the two key items. The revised DECLARE statement is:

```
declare hash bpck(dataset: 'bpstudy', ordered: 'yes');
```

Output 3.28 shows the results of executing Example 3.10 with the modified DECLARE statement. The observations in this revised version of HIGHSYSTOLIC are now in order by the values of PTID and EXAMDATE.

Output 3.28 PROC PRINT of HIGHSYSTOLIC

Obs	ptid	examdate	weight	systolic	diastolic
1	56KJR3D9	01/02/2013	142	147	82
2	56KJR3D9	04/03/2013	133	141	79
3	V8810YQ4	01/02/2013	182	154	89
4	V8810YQ4	02/06/2013	175	148	83
5	V8810YQ4	03/06/2013	170	147	79

Specifying a Key Value That Is Not in the Hash Object When Traversing the Hash Object

A hash object must always be defined with at least one key item. When you use the FIND method to find a specific key value and the key value is not in the hash object, SAS does not return any data from the hash object, and it sets a non-zero return code for the operation. Similarly, when you use the SETCUR method on a hash iterator object and specify a key value that is not present in the underlying hash object, SAS returns no data and it sets a non-zero return code for the operation. The hash iterator object does not move either to the next entry or to the previous entry in the hash object and return data from that entry.

The goal of Example 3.11 is to return data for patient 56KJR3D9 from the summer months of June, July, and August in 2013. The measurements were taken on the first Wednesday of each month. The SETCUR method call specifies the key value for PTID and it specifies an exam date key value of June 1, 2013. However, the first Wednesday in June 2013 is the 5th.

Example 3.11 Using SETCUR to Specify a Key Value That Is Not Present in the Hash Object

```
data summer;
  attrib ptid length=$8
         examdate format=mmddyy10.
         weight length=8
         systolic length=8
         diastolic length=8;

  declare hash bpck(dataset: 'bpstudy',ordered: 'yes');
  declare hiter bpckiter('bpck');
  bpck.definekey('ptid','examdate');
  bpck.definedata('ptid','examdate','weight','systolic','diastolic');
  bpck.definedone();

  call missing(ptid,examdate,weight,systolic,diastolic);

  drop rc;

  rc=bpckiter.setcur(key: '56KJR3D9',key: '01jun2013'd);
  put _all_;
  do while (rc=0);
    if month(examdate) gt 8 then stop;
    else output;
    rc=bpckiter.next();
  end;
run;
```

Even though patient 56KJR3D9 has measurements in June, July, and August, the DATA step does not write any observations to SUMMER. The SETCUR method does not find an entry in BPCK for June 1, 2013 so SAS assigns a non-zero value to variable RC. Therefore, the DO WHILE block does not execute since the value of RC is not zero when the DO WHILE statement executes.

The PUT _ALL_ statement following the SETCUR method lists the contents of the PDV after SETCUR executes. All of the DATA step variables contain missing values.

```
NOTE: There were 24 observations read from the data set WORK.BPSTUDY.
ptid=  examdate=. weight=. systolic=. diastolic=. rc=160038 _ERROR_=0
_N_=1
```

```
NOTE: The data set WORK.SUMMER has 0 observations and 6 variables.
```

If you modify the SETCUR method so that the key value supplied for EXAMDATE is June 5, 2013, Example 3.11 outputs the summer measurements for patient 56KJR3D9 to SUMMER.

```
rc=bpckiter.setcur(key: '56KJR3D9',key: '05jun2013'd);
```

Output 3.29 shows the observations in SUMMER that was created by the DATA step that includes the revised SETCUR method.

Output 3.29 PROC PRINT of SUMMER

Obs	ptid	examdate	weight	systolic	diastolic
1	56KJR3D9	06/05/2013	134	120	80
2	56KJR3D9	07/03/2013	133	124	81
3	56KJR3D9	08/07/2013	130	125	74

Application Example: Traversing a Hash Object from Beginning to End

Example 3.12 is a simple application that uses a hash iterator object to manage retrieval of data from a hash object. The DECLARE statement specifies the ORDERED: 'YES' argument tag so SAS retrieves data from the hash object in ascending order by the values of the key items. Example 3.12 creates two new variables from the data returned from the hash object.

Data set TESTGROUP contains dates of birth, test dates, and three treatment group assignments for 20 people. Example 3.12 has three goals:

- produce a data set that is in order by treatment group, date of birth, and ID
- compute the test subject's age
- recode the value of TREATMENT based on the test subject's age

With only 20 observations in TESTGROUP, which is the data set that Example 3.12 loads into the hash object, a processing advantage does not exist if you use a hash iterator object instead of a PROC SORT step followed by a DATA step with IF statements. However, when your data sets are larger, you are more likely to see efficiency gains if you use a hash iterator object and hash object because you would not have to sort or index the data set either before or after the DATA step.

Output 3.30 shows that the observations in TESTGROUP are not in any defined order.

Output 3.30 PROC PRINT of TESTGROUP

Obs	testid	dob	testdate	treatment
1	1014	02/07/1911	02/16/2013	Activity A
2	1008	04/07/1942	02/12/2013	Activity A
3	1003	12/10/1954	02/21/2013	Activity A
4	1007	02/22/1916	03/07/2013	Activity B
5	1020	07/03/1917	03/04/2013	Activity B
6	1010	06/02/1956	03/01/2013	Activity B
7	1016	12/05/1923	03/10/2013	No Activity
8	1006	04/22/1958	02/22/2013	Activity A
9	1009	02/23/1949	02/23/2013	Activity A
10	1017	12/20/1919	02/13/2013	Activity B
11	1013	09/05/1931	03/08/2013	No Activity
12	1001	12/21/1919	02/12/2013	No Activity
13	1012	07/09/1944	02/15/2013	No Activity
14	1015	03/20/1916	03/05/2013	No Activity
15	1018	02/23/1951	03/13/2013	Activity B
16	1004	05/26/1941	03/09/2013	Activity A
17	1011	03/19/1954	03/12/2013	Activity A
18	1019	04/06/1941	03/07/2013	No Activity
19	1005	06/23/1937	02/20/2013	Activity B
20	1002	06/03/1926	02/13/2013	Activity B

Example 3.12 loads data set TESTGROUP into hash object TG and specifies that retrieval from TG be in order by the values of TREATMENT, DOB, and TESTID. The second DECLARE statement associates hash iterator object TGI with hash object TG. All of the variables in TESTGROUP are named as data items on the DEFINEDATA method call.

The DATA step iterates once. The FIRST method call starts at the first entry in TG and returns its data to the DATA step. The DO UNTIL loop computes AGE and recodes TREATMENT for all entries retrieved from TG. On each iteration of the DO UNTIL loop, the NEXT method call attempts to read the next entry from TG and retrieve its four data items. When the NEXT method reaches the end of TG, SAS assigns a non-zero return code to the results of the call to NEXT and then exits the DO WHILE loop.

Example 3.12 Traversing a Hash Object from Beginning to End

```
data testcategories;
  attrib testid    length=8
         dob       length=8 format=mmddyy10.
         testdate  length=8 format=mmddyy10.
         age       length=8
         treatment length=$25;

  declare hash tg(dataset: 'testgroup', ordered: 'yes');
  declare hiter tgi('tg');
  tg.definekey('treatment','dob','testid');
  tg.definedata('testid','dob','testdate','treatment');
  tg.definedone();

  call missing(testid,dob,testdate,treatment);

  drop rc;

  rc=tgi.first();
  do while(rc eq 0);
    age=floor((testdate-dob)/365.25);
    if age lt 60 then treatment=catx(' ','Youngest:',treatment);
    else if 60 le age lt 80 then treatment=
                          catx(' ','Middle:',treatment);
    else if 80 le age lt 90 then treatment=
                          catx(' ','Oldest:',treatment);
    else if age ge 90 then treatment=catx(' ','Exempt:',treatment);
    output;
    rc=tgi.next();
  end;
run;
```

Output 3.31 shows that SAS has output the observations to data set TESTCATEGORIES in order by TREATMENT's original values, DOB, and TESTID.

Output 3.31 PROC PRINT of TESTCATEGORIES

Obs	testid	dob	testdate	age	treatment
1	1014	02/07/1911	02/16/2013	102	Exempt: Activity A
2	1004	05/26/1941	03/09/2013	71	Middle: Activity A
3	1008	04/07/1942	02/12/2013	70	Middle: Activity A
4	1009	02/23/1949	02/23/2013	64	Middle: Activity A
5	1011	03/19/1954	03/12/2013	58	Youngest: Activity A
6	1003	12/10/1954	02/21/2013	58	Youngest: Activity A
7	1006	04/22/1958	02/22/2013	54	Youngest: Activity A
8	1007	02/22/1916	03/07/2013	97	Exempt: Activity B

Obs	testid	dob	testdate	age	treatment
9	1020	07/03/1917	03/04/2013	95	Exempt: Activity B
10	1017	12/20/1919	02/13/2013	93	Exempt: Activity B
11	1002	06/03/1926	02/13/2013	86	Oldest: Activity B
12	1005	06/23/1937	02/20/2013	75	Middle: Activity B
13	1018	02/23/1951	03/13/2013	62	Middle: Activity B
14	1010	06/02/1956	03/01/2013	56	Youngest: Activity B
15	1015	03/20/1916	03/05/2013	96	Exempt: No Activity
16	1001	12/21/1919	02/12/2013	93	Exempt: No Activity
17	1016	12/05/1923	03/10/2013	89	Oldest: No Activity
18	1013	09/05/1931	03/08/2013	81	Oldest: No Activity
19	1019	04/06/1941	03/07/2013	71	Middle: No Activity
20	1012	07/09/1944	02/15/2013	68	Middle: No Activity

Application Example: Linking Hierarchically Related Data Sets

Example 3.13 shows how a hash object and hash iterator object can link two data sets that are hierarchically related and can summarize the data from the data set lower in the hierarchy.

The data used in Example 3.13 is from a survey where information was gathered about households and the people living in them. Data set HH stores the general information about the household. Data set PERSONS stores data about the persons living in the households. Variable HHID uniquely identifies each household, and this variable is common to both data sets. The two data sets are hierarchically related. Each household can have one or more people living in the household. Observations in PERSONS are uniquely identified by two variables: the household ID variable HHID, and the person ID variable PERSONID.

The goal of Example 3.13 is to create a data set with one observation per household. This observation should contain the general data obtained for a household in HH and a summarization of the person information in PERSONS for each household. The DATA step uses a hash object and hash iterator object to process the person records so that it can determine four statistics for each household:

- total number of people living in the household
- number of people in each of three age groups
- highest level of education in the household
- total income in the household

When the households were surveyed, the persons interviewed were assigned a sequential ID value starting with 1. All households have at least one household member.

Another way to summarize the person data in PERSONS is with PROC MEANS. You would need to follow the PROC MEANS step that saved the summary statistics in an output data set with either a DATA step or PROC SQL step. This subsequent step would match the output data set from PROC MEANS with the general household data found in HH.

Advantages of summarizing the data with hash objects in the DATA step is that you may only need one step and you can add complex programming statements to work with the summaries. However, if your data are not consistently defined as they are in Example 3.13, you would have to add additional programming statements.

Output 3.32 shows the contents of HH. Data set HH contains data for 10 households.

Output 3.32 PROC PRINT of HH

Obs	hhid	tract	surveydate	hhtype
1	HH01	CS	07/09/2012	Owner
2	HH02	CN	03/11/2012	Renter
3	HH03	CS	05/20/2012	Owner
4	HH04	SW	01/12/2012	Owner
5	HH05	NE	10/17/2012	Renter
6	HH06	NE	05/15/2012	Owner
7	HH07	SW	02/02/2012	Owner
8	HH08	NE	04/09/2012	Renter
9	HH09	CE	11/01/2012	Owner
10	HH10	CN	03/31/2012	Owner

Data set PERSONS contains at least one observation for every household in HH. Output 3.33 shows the contents of PERSONS.

Output 3.33 PROC PRINT of PERSONS

Obs	hhid	personid	age	gender	income	educlevel
1	HH01	P01	68	M	$52,000	12
2	HH01	P02	68	F	$23,000	12
3	HH02	P01	42	M	$168,100	22
4	HH03	P01	79	F	$38,000	10
5	HH04	P01	32	F	$56,000	16
6	HH04	P02	31	M	$72,000	18
7	HH04	P03	5	F	$0	0
8	HH04	P04	2	F	$0	0
9	HH05	P01	26	M	$89,000	22
10	HH06	P01	56	M	$123,000	18

Obs	hhid	personid	age	gender	income	educlevel
11	HH06	P02	48	F	$139,300	18
12	HH06	P03	17	F	$5,000	11
13	HH07	P01	48	M	$90,120	16
14	HH07	P02	50	F	$78,000	18
15	HH08	P01	59	F	$55,500	16
16	HH09	P01	32	F	$48,900	14
17	HH09	P02	10	M	$0	5
18	HH10	P01	47	F	$78,000	16
19	HH10	P02	22	F	$32,000	16
20	HH10	P03	19	M	$20,000	12
21	HH10	P04	14	M	$0	8
22	HH10	P05	12	F	$0	6
23	HH10	P06	9	F	$0	4

Example 3.13 starts by defining hash object P and hash iterator object PI. It loads all observations from PERSONS into P. Variable HHID is renamed to PHHID so that the HHID value retrieved from PERSONS can be compared to the value in DATA step variable HHID. The entries in hash object P are keyed and ordered by PHHID and PERSONID.

The SET statement reads observations from HH one observation at a time. The DO UNTIL loop iterates through hash object P and finds data for all persons in the household observation currently being processed. Each household has an observation in PERSONS where PERSONID= 'P01'. The SETCUR method call that precedes the DO UNTIL statement positions the search at the household's first person observation where PERSONID='P01'.

Statements in the DO UNTIL loop summarize the person information for the household. Since SAS retrieves data from P in order by PHHID and PERSONID, the NEXT method moves sequentially through the person records within each household. The DO UNTIL loop stops either when the values of HHID and PHHID are unequal or when the return code from the NEXT method is not zero. The latter situation occurs after SAS reads the last entry in P.

When the values of HHID and PHHID are unequal, it means that the current entry read from P belongs to the next household and that SAS has read all person entries for the previous household. The DO UNTIL loop ends and the current iteration of the DATA step ends. Control returns to the top of the DATA step and the SET statement reads the next household observation from HH. The DO UNTIL loop executes again at the first person entry for this next household in P.

Each iteration of the DATA step initializes the six household summary variables to 0. Variables NPERSONS, NPLT18, NP18_64, and NP65PLUS tally the total number of people in the household and the number in each of three age groups: less than 18; 18–64; and greater than 64. Variable HHINCOME tallies the income earned by all members of the household. Variable HIGHESTED is the maximum value of EDUCLEVEL in the household.

Example 3.13 Linking Two Hierarchically Related Data Sets

```
data hhsummary;
  attrib hhid length=$4
         tract length=$4
         surveydate length=8 format=mmddyy10.
         hhtype length=$10
         phhid length=$4
         personid length=$4
         age length=3
         gender length=$1
         income length=8 format=dollar12.
         educlevel length=3
         npersons length=3
         nplt18 length=3
         np18_64 length=3
         np65plus length=3
         highested length=3
         hhincome  length=8 format=dollar12.;

  if _n_=1 then do;
    declare hash p(dataset: 'persons(rename=(hhid=phhid))',ordered: 'yes');
    declare hiter pi('p');
    p.definekey('phhid','personid');
    p.definedata('phhid','personid','age','gender','income','educlevel');
    p.definedone();
    call missing(phhid,personid,age,gender,income,educlevel);
  end;

  keep hhid tract surveydate hhtype npersons nplt18 np18_64 np65plus
       highested hhincome;
  array zeroes{*} npersons nplt18 np18_64 np65plus hhincome highested;

  set hh;

  do i=1 to dim(zeroes);
    zeroes{i}=0;
  end;

  rc=pi.setcur(key: hhid, key: 'P01');
  do until(rc ne 0 or hhid ne phhid);
    if rc=0 then do;
      npersons+1;
      if age lt 18 then nplt18+1;
```

```
        else if 18 le age le 64 then np18_64+1;
        else if age ge 65 then np65plus+1;
        hhincome+income;
        if educlevel gt highested then highested=educlevel;
      end;
      rc=pi.next();
    end;
  run;
```

Output 3.34 shows the contents of HHSUMMARY. Each observation has general information for a household that came from data set HH, and a summary of the person information for each household that came from PERSONS.

Output 3.34 PROC PRINT of HHSUMMARY

Obs	hhid	tract	surveydate	hhtype	npersons	nplt18	np18_64	np65plus	highested	hhincome
1	HH01	CS	07/09/2012	Owner	2	0	0	2	12	$75,000
2	HH02	CN	03/11/2012	Renter	1	0	1	0	22	$168,100
3	HH03	CS	05/20/2012	Owner	1	0	0	1	10	$38,000
4	HH04	SW	01/12/2012	Owner	4	2	2	0	18	$128,000
5	HH05	NE	10/17/2012	Renter	1	0	1	0	22	$89,000
6	HH06	NE	05/15/2012	Owner	3	1	2	0	18	$267,300
7	HH07	SW	02/02/2012	Owner	2	0	2	0	18	$168,120
8	HH08	NE	04/09/2012	Renter	1	0	1	0	16	$55,500
9	HH09	CE	11/01/2012	Owner	2	1	1	0	14	$48,900
10	HH10	CN	03/31/2012	Owner	6	3	3	0	16	$130,000

You could revise Example 3.13 so that hash object P has one key, the HHID, and have it allow multiple sets of data items per key value. Methods like FIND_NEXT and FIND_PREV can traverse hash objects that allow multiple sets of data items per key value. The examples in Chapter 5 apply these techniques. Example 5.14 uses a hash object that allows multiple sets of data items per key value to produce the same data set as Example 3.13 does.

Application Example: Performing a Fuzzy Merge Using a Hash Iterator Object

Match-merging two or more data sets by the values of one or more variables is a common SAS programming task. The MERGE statement in the DATA step and joins in PROC SQL can match-merge data sets and tables. When you want to match observations by values that are not identical, a "fuzzy merge", you may need to recode your matching variables and add programming statements to your DATA step. The task is easier with PROC SQL because expressions can be written in the ON clause that specifies how to join two tables.

Example 3.14 shows how a DATA step with a hash iterator object can also perform a fuzzy merge. The goal is to find the prize to award to a contest participant based on the date the person completed a task. Different prizes are awarded during specific time periods.

Data set AWARDS shown in Output 3.35 contains the list of prizes to award during specific time periods. For example, participants who complete the task between February 1 and February 28 will receive a shopping certificate.

Output 3.35 PROC PRINT of AWARDS

Obs	awarddate	prize
1	01/31/2013	Restaurant Certificate
2	02/28/2013	Shopping Certificate
3	03/15/2013	Theater Tickets
4	04/15/2013	Baseball Tickets

Data set FINISHERS shown in Output 3.36 contains the list of people who will receive a prize based on the date they completed a task. Variable COMPLETED stores the completion date. Only one value of COMPLETED in FINISHERS, which is for Robert Jones in observation 11, exactly matches a value of AWARDDATE in AWARDS.

Output 3.36 PROC PRINT of FINISHERS

Obs	name	completed
1	Moore, Kathryn	12/27/2012
2	Jackson, Barbara	01/15/2013
3	Brown, Shannon	03/23/2013
4	Williams, Debra	03/26/2013
5	Harris, Joseph	02/01/2013
6	Brown, Patricia	01/08/2013
7	Johnson, Christopher	02/17/2013
8	Rodriguez, Shawn	03/31/2013
9	Gonzalez, Patrick	01/14/2013
10	Wright, Nicholas	03/02/2013
11	Jones, Robert	02/28/2013
12	Miller, Christopher	03/25/2013
13	Rogers, Francie	05/15/2013

By comparing the values of AWARDDATE and COMPLETED, Example 3.14 determines the prize awarded to a participant if the participant's completion date is on or before the award date.

Example 3.14 reads data set FINISHERS. It loads data set AWARDS into hash object AZ and associates hash iterator object AZI with AZ. Hash object AZ has one key, AWARDDATE. The

ORDERED: YES argument specifies that SAS retrieve the values of AWARDDATE from AZ in chronological order.

For each observation read from FINISHERS, the DATA step starts at the beginning of AZ by applying the FIRST method to AZI. The DO UNTIL loop compares the value of COMPLETED to AWARDDATE. The NEXT method reads the successive entries in AZ. The prize is selected when the value of COMPLETED is less than or equal to the value of AWARDDATE. Variable FOUNDMATCH indicates whether a match has been found. If SAS does not find a match in AZ, it assigns default values to AWARDDATE and PRIZE. The last observation has a completion date of May 15, which is after the latest date of April 30 in AWARDS. SAS assigns the default values of AWARDDATE and PRIZE to this observation.

Example 3.14 Performing a Fuzzy Merge Using a Hash Iterator Object

```
data award_list;
  attrib completed length=8 format=mmddyy10.
         awarddate length=8 format=mmddyy10.
         prize     length=$22;

  if _n_=1 then do;
    declare hash az(dataset: 'awards', ordered: 'yes');
    declare hiter azi('az');
    az.definekey('awarddate');
    az.definedata('awarddate','prize');
    az.definedone();
    call missing(awarddate,prize);
  end;
  set finishers;

  drop rc foundmatch;

  foundmatch=0;
  rc=azi.first();
  do until(rc ne 0);
    if completed le awarddate then do;
      foundmatch=1;
      leave;
    end;
    rc=azi.next();
  end;
  if not foundmatch then do;
    awarddate=.;
    prize='** Post contest';
  end;
run;
```

Output 3.37 shows the contents of AWARD_LIST.

Output 3.37 PROC PRINT of AWARD_LIST

Obs	completed	awarddate	prize	name
1	12/27/2012	01/31/2013	Restaurant Certificate	Moore, Kathryn
2	01/15/2013	01/31/2013	Restaurant Certificate	Jackson, Barbara
3	03/23/2013	04/15/2013	Baseball Tickets	Brown, Shannon
4	03/26/2013	04/15/2013	Baseball Tickets	Williams, Debra
5	02/01/2013	02/28/2013	Shopping Certificate	Harris, Joseph
6	01/08/2013	01/31/2013	Restaurant Certificate	Brown, Patricia
7	02/17/2013	02/28/2013	Shopping Certificate	Johnson, Christopher
8	03/31/2013	04/15/2013	Baseball Tickets	Rodriguez, Shawn
9	01/14/2013	01/31/2013	Restaurant Certificate	Gonzalez, Patrick
10	03/02/2013	03/15/2013	Theater Tickets	Wright, Nicholas
11	02/28/2013	02/28/2013	Shopping Certificate	Jones, Robert
12	03/25/2013	04/15/2013	Baseball Tickets	Miller, Christopher
13	05/15/2013	.	** Post contest	Rogers, Francie

Chapter 4: Creating Data Sets from Hash Objects and Updating Contents of Hash Objects

The examples in Chapter 3 created hash objects from existing data sets. They loaded data from data sets into hash objects, and they retrieved information from them during execution of the DATA step. The information in the hash object was not modified in any way.

This chapter describes how to work more dynamically with hash objects. The examples demonstrate how to use methods that can create data sets from hash objects and how to apply methods that can add, modify, or remove data from hash objects.

The code associated with the hash objects in this chapter assumes that the hash objects have only one set of data items per key value. This is the default when you declare a hash object. It is possible to allow multiple sets of data items per key value in a hash object. Chapter 5 presents examples that use hash objects that allow multiple sets of data items per key value.

Creating a Data Set from a Hash Object

A DATA step's purpose is to process data and many times create a new data set. You start the DATA step with a DATA statement that names the data set that you want to create.

You can also create data sets from the contents of a hash object in a DATA step. You do this by applying the OUTPUT method and naming the data set that you want to create from the hash object on the OUTPUT method's DATASET argument tag. You do not name the data set on the DATA statement.

An advantage of being able to create data sets from hash objects with the OUTPUT method is that you can name your data sets during execution of your DATA steps. SAS language statements that call the OUTPUT method within a DATA step can name your data sets, and you can create as many data sets as your data and code require. You do not need to explicitly name the data sets that the OUTPUT method creates as the DATA statement requires you to do.

The syntax of the OUTPUT method follows. The OUTPUT method requires that you specify at least one DATASET argument tag.

```
rc=object.OUTPUT(DATASET: 'dataset-1 <(datasetoption)>'
<..., DATASET:  'dataset-n'>('datasetoption<(datasetoption)>');
```

When the OUTPUT method executes, SAS outputs the entire contents of the hash object to the data set. The OUTPUT method is not related to the OUTPUT SAS language statement where the OUTPUT statement must execute once for every observation you want SAS to add to a data set. Instead, the statement that calls the OUTPUT method typically executes just once during execution of your DATA step. All the information that you want written to your data set must already exist in your hash object before the DATA step executes the OUTPUT method.

Additionally, you can apply data set options to the data sets specified on the DATASET argument tags. For example, starting with SAS 9.3, you can control which variables SAS writes to a data set by adding either the KEEP= or DROP= data set options. Also, with the WHERE= data set option, you can control which observations the OUTPUT method writes to the data set it creates.

You can create more than one data set from one hash object with one application of the OUTPUT method. With the use of data set options, you can save one group of variables in one data set and save a second group of variables in a second data set as the following statement illustrates.

```
rc=myhash.output(dataset: 'demog(keep=ptid ptln ptfn ptmi ptdob ptgender)',
          dataset: 'insurance(keep=ptid ptinsurance1 ptinsurance2 )');
```

The OUTPUT method does not have any options that can prevent the replacement of an existing data set whose name you have specified on the DATASET argument tag. Further, SAS does not issue any warning messages that inform you that the data set named in the OUTPUT method call already exists.

SAS generates an error for the OUTPUT method if you specify the same data set name on both the DATA statement and the DATASET argument tag of the OUTPUT method. The DATA statement takes precedence over the OUTPUT method. Therefore, the OUTPUT method does not replace or contribute observations to the data set named on the DATA statement because it cannot open the data set.

Similarly, SAS generates an error for the OUTPUT method if your DATA step reads a data set that you also name on the DATASET argument tag of the OUTPUT method. SAS does not replace the data set when the OUTPUT method executes because the OUTPUT method cannot close a data set while SAS executes a SET or MERGE statement on a data set with the same name.

Adding, Modifying, and Removing Data from a Hash Object

A hash object is not limited to being a static object in which you look up data. Your SAS code can dynamically change the contents of your hash object during execution of your DATA step. Methods exist that can add, remove, and modify data in a hash object. You use these methods when you need to alter the contents of your hash object.

The syntax for the three main methods that add, modify, and remove data from a hash object follows:

```
rc=object.ADD(<KEY: keyvalue-1,..., KEY: keyvalue-n, DATA: datavalue-1,
          ..., DATA: datavalue-n>);
rc=object.REPLACE(<KEY: keyvalue-1,..., KEY: keyvalue-n,DATA:
              datavalue-1,..., DATA: datavalue-n>);
rc=object.REMOVE(<KEY: keyvalue-1,..., KEY: keyvalue-n>);
```

When passing key values to the hash object, either you can use the DATA step variable that has the same name as your key item or you can specify the key value with the KEY: argument tag. You can write your key value as a literal value, as another variable, or as a result of SAS expressions or functions.

Similarly, when passing data values to the hash object, you can use the DATA step variable that has the same name as your data item or you can specify the data value with the DATA: argument tag. Data values are specified in the same way that you specify key values: they can be specified as a literal value, as another variable, or as a result of SAS functions.

If you need to specify a key value or a data value as a literal, another variable, or a result of SAS functions, then you must specify all other items in the method call. This is true even if all other items have the same names and attributes as their corresponding DATA step variables. The sections and examples that follow further illustrate this concept.

The REF method can also add data to a hash object. This method consolidates the CHECK and ADD methods into one method call. The REF method is useful when your data contains multiple sets of data items for a key value, and you only need to process one set of data items. The REF method adds only one entry for a key value to the hash object.

```
rc=object.REF(<KEY: keyvalue-1,..., KEY: keyvalue-n, DATA: datavalue-1,
                 ..., DATA: datavalue-n>);
```

To illustrate an application of the REF method, a patient may have multiple appointments within a year, but you only need to determine whether a patient had *any* appointment. The patient's ID is your hash object's key item, and your hash object needs only one set of data items for each patient. The hash object does not need to contain an entry for each appointment that the patient had. The first time your DATA step reads an observation for a patient's appointment within the year, the REF method adds an entry to the hash object. SAS does not add any additional sets of data items to the hash object for any subsequent appointment observations it reads for the patient. Example 4.7 presents an example similar to this illustration.

Chapter 5 describes the techniques that allow you to define hash objects that allow multiple sets of data items per key value. In such a situation, a DATA step processing the data set described in the previous paragraph could define a hash object that is keyed by the patient's ID and contain a separate entry (or set of data items) for each appointment for a patient.

Defining the Key Structure in a Hash Object

The default structure of a hash object allows only unique key values. This means that you can have only one set of data items per unique combination of key values. Your hash objects can have one or more key items, and their values define the unique combinations in your data. Additionally, you can specify options on the DECLARE statement that allow multiple sets of data items per unique combination of key values. Chapter 5 describes these techniques.

The ADD, REMOVE, REPLACE, and REF methods behave differently depending on whether your code allows multiple sets of data items per key value. The examples in this chapter show you how to use these methods only under the condition of one set of data items per key value.

When you define hash objects that allow multiple sets of data items per key value, the two methods, REMOVEDUP and REPLACEDUP, when compared to REMOVE and REPLACE, give you more control in modifying the contents of your hash object. See Chapter 5 for more information.

Understanding How to Specify the KEY and DATA Argument Tags

When a key item or data item does not share the same name and attributes with a DATA step variable, you must specify KEY: and DATA: argument tags for the item. Further, in this situation, you must specify argument tags for all items in your hash object when you write your method call even if some of the items share the same names and attributes with DATA step variables. The KEY: and DATA: argument tags must be written in the same order that you defined them with the DEFINEKEY and DEFINEDATA methods.

When you create a data set from a hash object and your hash object has data items, you must also define your key items as data items if you want SAS to add the key items to the data set it creates with the OUTPUT method. SAS stores key items in one place and data items in another place. It outputs only the items defined as data. SAS uses the key items to manage storage in and retrieval from the hash object.

Example 4.1 illustrates the concepts of this section. The bulleted items are explained further following the DATA step.

Example 4.1 creates hash object CODELIST, but it does not load a data set into CODELIST. The DATA step iterates once and adds data to the hash object as it executes. The last action it takes is to apply the OUTPUT method to output all of the data in CODELIST to data set TESTCODES1.

Hash object CODELIST is keyed by CODE. The ORDERED: "YES" argument specified in the DECLARE statement retrieves data items in CODELIST in order by the key values of CODE.

Example 4.1 Illustrating the Usage of the KEY and DATA Argument Tags

```
data _null_;
  attrib code length=$6
         codedesc length=$25
         codedate length=8 format=mmddyy10.;

  declare hash codelist(ordered: 'yes');
  codelist.definekey('code');
  codelist.definedata('code','codedesc','codedate'); ❶
  codelist.definedone();
  call missing(code,codedesc,codedate);

  code='AB234Z';
  codedesc='Credit';    ❷
  codedate='05jan2013'd;
```

```
  rc=codelist.add();

  /* This statement is incomplete and stops the DATA step */ ❸
  ***rc=codelist.add(key: 'JK987B');

  /* This statement is complete */ ❹
  rc=codelist.add(key: 'JK987B', data: 'JK987B', data:codedesc,
                      data:codedate);

  /* This statement is complete */
  code='CU824P';
  rc=codelist.add(key: 'ZZ521L', data: code, ❺
      data:codedesc, data: '05mar2013'd);

  codefront='RS';
  codeback='438E';
  rc=codelist.add(key: cats(codefront,codeback),
                data: cats(codefront,codeback), ❻
                data: 'Debit', data: '23feb2013'd);

  rc=codelist.output(dataset: 'testcodes1');
run;
```

❶ Note that the key variable CODE is also a data item. If you did not also specify CODE as a data item, the OUTPUT method would write only two variables, CODEDESC and CODEDATE, to data set TESTCODES1.

❷ SAS fills the first entry in CODELIST with the data defined by the first three assignment statements.

❸ If statement `rc=codelist.add(key: 'JK987B')` was not commented out, the DATA step would not execute. Even though variables CODEDESC and CODEDATE have values, SAS requires that you specify argument tags for *all* of the key and data items if you need to specify *any* of them.

❹ This call to the ADD method is the complete version of what the call to ADD in #3 was trying to accomplish. Here all the key items and data items are specified. The data items are specified in the order that the DEFINEDATA method defined them.

❺ Note that the values for the CODE key item and CODE data item are different on this call to the ADD method. The value for the CODE *key item* is explicitly defined as 'ZZ521L'. The value for the CODE *DATA step variable* is 'CU824P'. This latter value, 'CU824P', is the value that SAS adds to the hash object as the CODE *data* value. Because the DECLARE statement specified that the items be ordered by the key values of CODE, the observation corresponding to this entry where the *key* value is 'ZZ521L' is last in the PROC PRINT output. This causes the observations in TESTCODES to not be in order by the variable CODE as you might expect. Output 4.1 shows the order of the observations in TESTCODES1. These results illustrate that SAS stores hash key items and hash data items separately.

❻ This call to the ADD method shows that you can apply functions when specifying values to argument tags. Here the key value and data value for CODE are each the concatenation of two variables.

Output 4.1 lists the observations in data set TESTCODES1, which the OUTPUT method call in Example 4.1 creates.

Output 4.1 PROC PRINT of TESTCODES1

Obs	code	codedesc	codedate
1	AB234Z	Credit	01/05/2013
2	JK987B	Credit	01/05/2013
3	RS438E	Debit	02/23/2013
4	CU824P	Credit	03/05/2013

Identifying the Variables That the OUTPUT Method Writes to a Data Set

When you create a data set from a hash object with the OUTPUT method, SAS writes only the variables named in the DEFINEDATA method to the data sets specified on the DATASET argument tags in the OUTPUT method. SAS does not output the variables you list in the DEFINEKEY statement unless these variables are also listed in the DEFINEDATA method. Therefore, when you define the hash object that you want to save in a data set, list in the DEFINEDATA method all of the variables that you want written to that data set.

In Example 4.1, both the DEFINEKEY and DEFINEDATA methods specified item CODE. When the OUTPUT method executes, SAS creates data set TESTCODES1 and includes variable CODE in the data set.

The DATA step in Example 4.2 is similar to the DATA step in Example 4.1. Example 4.2 shows what happens when you do ***not*** specify CODE on the DEFINEDATA statement. While CODE remains a key of hash object CODELIST, SAS does not add this variable to data set TESTCODES2. The DATA step writes two entries to hash object CODELIST, and it writes two observations and two variables to data set TESTCODES2.

Since the DECLARE statement includes the ORDERED: "Y" argument tag, SAS retrieves entries from hash object CODELIST in order by the values of CODE.

Example 4.2 Identifying the Variables That the OUTPUT Method Writes to a Data Set

```
data _null_;
  attrib code length=$6
         codedesc length=$25
         codedate length=8 format=mmddyy10.;
```

```
    declare hash codelist(ordered: 'y');
    codelist.definekey('code');
    codelist.definedata('codedesc','codedate');
    codelist.definedone();

    call missing(code,codedesc,codedate);

    code='CU824P';
    codedesc='Credit';
    codedate='05jan2013'd;
    rc=codelist.add();

    rc=codelist.add(key: 'AB234Z',
                    data:codedesc, data: '05mar2013'd);

    rc=codelist.output(dataset: 'testcodes2');
  run;
```

Output 4.2 shows the contents of TESTCODES2.

Output 4.2 PROC PRINT of TESTCODES2

Obs	codedesc	codedate
1	Credit	03/05/2013
2	Credit	01/05/2013

Understanding the Interaction between DATA Step Variables and Hash Object Data Items When Replacing Data in a Hash Object

When SAS compiles a DATA step, it creates the Program Data Vector (PDV). The PDV contains all of the variables in the input data set, the variables created in DATA step statements, and the two automatic variables, _N_ and _ERROR_. SAS uses the PDV as a place from which to transfer data between your DATA step variables and hash object data items. The interaction between DATA step variables and hash object data items is important to understand when you work with methods that modify the contents of your hash objects.

When SAS retrieves data from a hash object with the FIND method, SAS takes data items from the hash object and overwrites the values of the same-named variables in the PDV for the observation that it is currently processing. Therefore, if you need to compare a value in the hash object with one in the observation that SAS is currently processing, you must rename the data item in the hash object or rename the DATA step variable. This new variable prevents your code from overwriting the DATA step variable's values so that your code can make its comparisons.

The DATA steps in Examples 3.1 and 3.2 and PROC PRINT outputs in Outputs 3.2 and 3.3 illustrate how SAS handles DATA step variables and hash object data items with the same name when it applies the FIND method and CHECK method. Example 3.1 demonstrated that the data values retrieved from a hash object with the FIND method replace the values in the DATA step variables that share the same names as the hash object items. Example 3.2 used the CHECK method to show that the CHECK method finds the presence of a key value and that it does not return any data values to the DATA step.

Using the same data set and general process as Examples 3.1 and 3.2, Example 4.3 demonstrates the REPLACE method. Example 4.3 loads data set CODELOOKUP into hash object HASHSOURCE. Data set CODELOOKUP has two observations and three variables. The second observation in CODELOOKUP has missing values for variables CODEDESC and CODEDATE. Output 4.3 lists the two observations in data set CODELOOKUP.

Output 4.3 Data Set CODELOOKUP

Obs	code	codedesc	codedate
1	AB871K	Credit	02/08/2013
2	GG401P		.

Example 4.3 starts with an ATTRIB statement that defines DATA step variables that correspond to the key and data items in hash object HASHSOURCE. Next, SAS loads data set CODELOOKUP into HASHSOURCE. Example 4.3 does not read a data set with a SET statement or any other data reading statement. The DATA step instead creates the observations.

The first two times Example 4.3 calls the REPLACE method, SAS updates the contents of hash object HASHSOURCE for the two entries where CODE='AB871K' and CODE='GG401P' because these entries already exist in HASHSOURCE. The third call to the REPLACE method adds a new entry to HASHSOURCE since HASHSOURCE does not contain an entry for CODE='MI497Q'.

The final statement applies the OUTPUT method to create data set NEWCODELOOKUP. This data set contains the updated contents of HASHSOURCE.

The three OUTPUT statements in the DATA step write observations to data set UPDATES3. The contents of data set UPDATES3 and NEWCODELOOKUP are identical except that data set UPDATES3 includes variable RC. Outputs 4.4 and 4.5 list the observations in these two data sets.

Example 4.3 Illustrating How the REPLACE Method Updates Data in a Hash Object

```
data updates3;
  attrib code length=$6
         codedesc length=$25
         codedate length=8 format=mmddyy10.;

  declare hash hashsource(dataset: 'codelookup',ordered: 'y');
  hashsource.definekey('code');
  hashsource.definedata('code','codedesc','codedate');
  hashsource.definedone();
```

```
    call missing(code,codedesc,codedate);

    code='AB871K';
    codedesc='Debit';
    codedate='15feb2013'd;
    rc=hashsource.replace();
    output;
    code='GG401P';
    codedesc='Credit';
    codedate='23feb2013'd;
    rc=hashsource.replace();
    output;
    code='MI497Q';
    codedesc='Debit';
    codedate='03feb2013'd;
    rc=hashsource.replace();
    output;
    rc=hashsource.output(dataset: 'newcodelookup');
run;
```

Output 4.4 lists the observations in data set UPDATES3 that the OUTPUT statements create in Example 4.3.

Output 4.4 PROC PRINT of UPDATES3

Obs	code	codedesc	codedate	rc
1	AB871K	Debit	02/15/2013	0
2	GG401P	Credit	02/23/2013	0
3	MI497Q	Debit	02/03/2013	0

Output 4.5 lists data set NEWCODELOOKUP, which the OUTPUT method in the last statement of Example 4.4 creates.

Output 4.5 PROC PRINT of NEWCODELOOKUP

Obs	code	codedesc	codedate
1	AB871K	Debit	02/15/2013
2	GG401P	Credit	02/23/2013
3	MI497Q	Debit	02/03/2013

Replacing Key Item Values When a Key Item Is Also a Data Item

The purpose of the REPLACE method is to replace *data* item values in a hash object. This does not include the replacement of key item values. When SAS creates a hash object, it stores the key items separately from the data items. SAS uses the key items to plan the structure of your hash object so that it can efficiently access the data items.

You can add the name of your key item on the DEFINEDATA method. It is necessary to do so if you want the data set created by the OUTPUT method to contain a variable corresponding to the key item. However, if your REPLACE method call modifies the value of this data item, it does not also modify the same-named key item. The entries in your hash object continue to have the same set of key values that it started with. Therefore, the retrieval order of the entries in the hash object does not change.

It takes two steps to modify the value of a key item. First you must remove the entry from the hash object, and next you add back to the hash object the entry with its new key value. SAS revises its key storage structure to add this new key value, and retrieves entries from the hash object in order of the revised set of key values.

Examples 4.4 and 4.5 illustrate these concepts. Both examples load data set CODES into hash object CD. Data set CODES has three variables and three observations. Variable CODE is the key in CD. The ORDERED: "YES" argument specifies that SAS retrieve entries from CD in order by the values of CODE. Output 4.6 lists the observations in data set CODES.

Output 4.6 PROC PRINT of CODES

Obs	code	codedesc	codedate
1	AB871K	Debit	02/15/2013
2	GG401P	Credit	02/23/2013
3	MI497Q	Debit	02/03/2013

The goal in both Examples 4.4 and 4.5 is to modify CODE and CODEDESC in the third observation. The value of CODE should be changed to "C1UE23", and the value of CODEDESC should be changed to "NewAcct". You might expect that this new value of CODE should cause observation 2 to swap positions with observation 3 in the output data set. This swap in position does not happen in Example 4.4, and it does happen in Example 4.5.

Example 4.4 names CODE on both the DEFINEKEY and DEFINEDATA methods. It is necessary to name CODE on the DEFINEDATA method so that SAS includes CODE in the data set created by the OUTPUT method.

The FIND method finds the entry in CD where CODE="MI497Q", and SAS retrieves the data associated with this value of CODE. The FIND method sets the value of the CODE DATA step variable to "MI497Q".

Next, the REPLACE method modifies the entry where the key value is "MI497Q" by placing DATA step variable CODE as the object of the KEY argument tag. Since the goal is to change the data value of CODE="MI497Q" in hash object CD to "C1UE23", the value "C1UE23" is specified on the first DATA: argument tag as a literal. The object of the second DATA argument tag is the revised value of CODEDESC. Since it is not necessary to change the data item value of CODEDATE, the object of the third DATA argument tag is variable CODEDATE. The value for CODEDATE was obtained from CD by the FIND method.

The OUTPUT method writes the updated version of CD to data set NOKEYCHANGE. Output 4.7 shows that the values of CODE and CODEDESC have changed for the observation that started out as MI497Q, but that the order of the observations in NOKEYCHANGE remains the same as it was in input data set CODES.

Example 4.4 Attempting to Replace a Key Value for a Key Item That Is Also Named As a Data Item

```
data _null_;
  attrib code length=$6
         codedesc length=$25
         codedate length=8 format=mmddyy10.;

  declare hash cd(dataset: 'codes', ordered: 'yes');
  cd.definekey('code');
  cd.definedata('code','codedesc','codedate');
  cd.definedone();

  call missing(code,codedesc,codedate);

  rc=cd.find(key: 'MI497Q');
  rc=cd.replace(key: code, data: 'C1UE23', data: 'NewAcct',
                data: codedate);

  rc=cd.output(dataset: 'nokeychange');
run;
```

Output 4.7 lists the observations in NOKEYCHANGE.

Output 4.7 PROC PRINT of NOKEYCHANGE

Obs	code	codedesc	codedate
1	AB871K	Debit	02/15/2013
2	GG401P	Credit	02/23/2013
3	C1UE23	NewAcct	02/03/2013

Example 4.5 modifies Example 4.4 so that it repeats the same FIND method call, but its second step removes the entry it finds from the hash object with the REMOVE method. Third, it applies the ADD method to add a new entry to the hash object that has the updated values. This new entry causes SAS to revise the key structure of CD. Although the entry with CODE="C1UE23" was added last, SAS retrieves it before it retrieves the entry where CODE="GG401P".

An assignment statement modifies the DATA step variable CODE to equal the new value of C1UE23. This variable is the object of both the KEY argument tag and the first DATA argument tag in the ADD method call.

The value of CODEDATE remains the same as that retrieved by the FIND method so variable CODEDATE is the object of the third DATA argument tag. Output 4.8 shows that the observations in KEYCHANGED that the OUTPUT method creates from hash object CD are in order by the values of CODE.

Example 4.5 Replacing the Value of a Key Item That Is Also Named As a Data Item

```
data _null_;
  attrib code length=$6
         codedesc length=$25
         codedate length=8 format=mmddyy10.;

  declare hash cd(dataset: 'codes',ordered: 'y');
  cd.definekey('code');
  cd.definedata('code','codedesc','codedate');
  cd.definedone();

  call missing(code,codedesc,codedate);

  rc=cd.find(key: 'MI497Q');
  rc=cd.remove();
  code='C1UE23';
  rc=cd.add(key: code, data: code, data: 'NewAcct', data: codedate);

  rc=cd.output(dataset: 'keychanged');
run;
```

Output 4.8 lists the observations in KEYCHANGED.

Output 4.8 PROC PRINT of KEYCHANGED

Obs	code	codedesc	codedate
1	AB871K	Debit	02/15/2013
2	C1UE23	NewAcct	02/03/2013
3	GG401P	Credit	02/23/2013

Comparing the DATA Statement and the OUTPUT Method

Both the DATA statement and the OUTPUT method create data sets. How they create data sets differs. This section compares the two ways of creating data sets.

The DATA statement creates the data sets that it names. The OUTPUT method creates data sets that are specified on its DATASET argument tags. A single DATA step can create data sets with the DATA statement and also with statements that call the OUTPUT method. However, SAS does not successfully execute a DATA step where the data set name on the DATA statement is identical to the data set name supplied to the argument tag of the OUTPUT method.

A DATA step writes an observation to a data set named on the DATA statement every time an OUTPUT statement (or an implied OUTPUT statement) executes. When the OUTPUT method executes, SAS writes the contents currently in the hash object to the data sets specified on the DATASET argument tags. You would typically execute the OUTPUT method only once. It is not necessary to execute the OUTPUT method on each iteration of a DATA step. While executing the OUTPUT method on each iteration of the DATA step would not generate any errors, it could be very inefficient.

The DATA step in Example 4.6 specifies data set RANDOM10 on both the DATA statement and in the OUTPUT method call. The goal of Example 4.6 is to generate 10 random numbers between 1 and 100 using the UNIFORM function.

(Since the DATA step includes CALL STREAMINIT, SAS generates the same series of 10 numbers each time the DATA step executes. Without CALL STREAMINIT, SAS uses the clock time as its seed, and the set of 10 numbers that Example 4.6 generates each time it executes is not the same.)

Example 4.6 Illustrating the Problem of Specifying the Same Data Set on the DATA Statement and OUTPUT Method DATASET Argument Tag

```
data random10;
  attrib sequence length=8 label='Selection Order'
         random_number length=8 label='Random Number';

  declare hash r();
  r.definekey('sequence');
  r.definedata('sequence','random_number');
  r.definedone();

  call missing(sequence,random_number);
  call streaminit(12345);

  do sequence=1 to 10;
    random_number=ceil(100*rand('uniform'));
    rc=r.add();
    output;
  end;
  rc=r.output(dataset: 'random10');
run;
```

The DATA step in Example 4.6 does create a data set called RANDOM10, but it ends in error when it encounters the statement that calls the OUTPUT method. By the time the OUTPUT method executes, the DATA statement has already opened data set RANDOM10 and SAS cannot add the contents of hash object R to RANDOM10. The OUTPUT statement in the iterative DO loop successfully outputs 10 observations to RANDOM10.

SAS generates the following error message and note.

```
ERROR: Data set WORK.RANDOM10 is already open for output.

NOTE: The data set WORK.RANDOM10 has 10 observations and 3 variables.
```

If you replace the DATA RANDOM10 statement with a DATA _NULL_ statement, SAS successfully creates data set RANDOM10 with the OUTPUT method. Output 4.9 shows a PROC PRINT of data set RANDOM10 that the OUTPUT method call created when a DATA _NULL_ statement starts the DATA step.

Output 4.9 PROC PRINT of RANDOM10

Obs	sequence	random_number
1	2	89
2	5	66
3	1	86
4	9	49
5	3	84
6	7	54
7	4	12
8	10	51
9	6	20
10	8	16

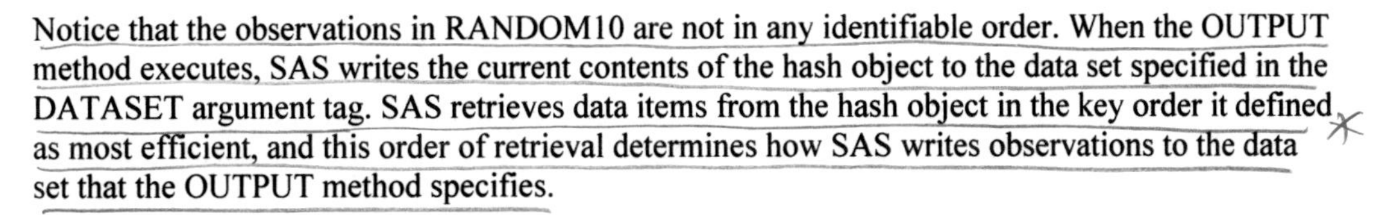

Notice that the observations in RANDOM10 are not in any identifiable order. When the OUTPUT method executes, SAS writes the current contents of the hash object to the data set specified in the DATASET argument tag. SAS retrieves data items from the hash object in the key order it defined as most efficient, and this order of retrieval determines how SAS writes observations to the data set that the OUTPUT method specifies.

You might expect SAS to arrange items in R in the order they were added. The DO loop executes sequentially so that the values of SEQUENCE go from 1 to 10, and the ADD method is applied in each iteration of the DO loop. However, unless you use the ORDERED: argument tag when you define your hash object, you will not know the order in which SAS retrieves items from the hash object.

Wisely using the ORDERED argument tag and selecting keys when you output a hash object to a data set with the OUTPUT method can save resources that otherwise would be used if a PROC SORT step followed your DATA step.

If you add the ORDERED argument tag to the DECLARE statement as follows, when the OUTPUT method executes, it writes the contents of R to data set RANDOM10 in ascending order by the values of key item SEQUENCE.

```
declare hash r(ordered: 'yes');
```

Output 4.10 shows the contents of RANDOM10 when the ORDERED argument tag is added to the DECLARE statement.

Output 4.10 PROC PRINT of RANDOM10 after Specifying the ORDERED: "YES" Argument Tag

Obs	sequence	random_number
1	1	86
2	2	89
3	3	84
4	4	12
5	5	66
6	6	20
7	7	54
8	8	16
9	9	49
10	10	51

Using the ORDERED argument tag gives you added flexibility within the DATA step to arrange the observations in the data set you create with the OUTPUT method. For example, by changing the option to 'd' or 'descending', you can reverse the order of the observations so that the last selected random number is the first observation in RANDOM10.

```
declare hash r(ordered: 'descending');
```

Similarly, changing the key items in conjunction with the ORDERED argument can also alter the order of the observations in the data set you create with the OUTPUT method. For example, if you change the key item from SEQUENCE to RANDOM_NUMBER, SAS outputs observations to RANDOM10 in order by the ascending values of RANDOM_NUMBER.

```
r.definekey('random_number');
```

Output 4.11 shows RANDOM10 when the key item is RANDOM_NUMBER and the option on the ORDERED argument tag is 'YES'.

Output 4.11 Data Set RANDOM10 When the Key Item Is RANDOM_NUMBER and ORDERED: "YES" Is Specified

Obs	sequence	random_number
1	4	12
2	8	16
3	6	20
4	9	49
5	10	51
6	7	54
7	5	66
8	3	84
9	1	86
10	2	89

Application Example: Finding the Unique Values of a Variable in a Data Set

The goal of Example 4.7 is to collect the unique values of a variable in a data set. This task can be easily accomplished with BY variable processing in the DATA step, with PROC FREQ, or with the DISTINCT function in PROC SQL. This example shows how to obtain the same outcome in one DATA step with a hash object.

A hash object solution for finding the unique values is advantageous when the data set that you have to examine has many observations. Such a solution can reduce processing time and resource usage. Advantages of using a hash object include:

- Your input data set does not have to be sorted or indexed prior to execution of the DATA step as it would with BY-variable processing in the DATA step. PROC SQL may not process a large table efficiently. Its efficiency would improve if you sort or index the table prior to execution of the PROC SQL step.
- You can retrieve data from your hash object in order by the values of the key items so that when SAS outputs the hash object, it writes the observations in sorted order.
- You can save other information in the hash object besides the unique values. If you used PROC FREQ, you would only be able to save the set of unique values and the statistics that PROC FREQ calculates.

Input data set CLAIMS2013 in the following DATA step contains several thousand observations of medical claims data. Each observation contains data for one medical claim for one patient on one date. Variable PROVIDER_ID contains the ID for the place that provided care to the patient. The goal is to identify the set of unique provider ID values in CLAIMS2013, and output that set of IDs in sorted order to a data set.

Output 4.12 lists the first 20 observations in CLAIMS2013.

Output 4.12 PROC PRINT of CLAIMS2013 (first 20 observations)

Obs	ptid	claimdate	provider_id	charge	provider_type
1	4DGNWU3Z	04/12/2013	OPSURGERY039	$7,902.00	OPSURGERY
2	4DGNWU3Z	02/07/2013	PHYSTHERAPY071	$330.40	PHYSTHERAPY
3	XZRQTXGU	04/17/2013	OPRADIOLOGY069	$1,517.60	OPRADIOLOGY
4	JT1Z17GE	08/22/2013	CLINIC016	$53.00	CLINIC
5	THSALN2J	09/25/2013	OPRADIOLOGY070	$450.20	OPRADIOLOGY
6	THSALN2J	11/11/2013	CLINIC014	$229.80	CLINIC
7	THSALN2J	08/15/2013	CLINIC010	$102.20	CLINIC
8	THSALN2J	09/09/2013	LAB049	$147.40	LAB
9	DWJSEGFQ	07/31/2013	PHYSTHERAPY074	$381.40	PHYSTHERAPY
10	DWJSEGFQ	10/04/2013	CLINIC023	$65.80	CLINIC
11	L2PAT59T	10/16/2013	OPSURGERY044	$3,373.80	OPSURGERY
12	GE356WUG	10/07/2013	OPSURGERY035	$7,128.20	OPSURGERY
13	GE356WUG	07/24/2013	CLINIC005	$104.60	CLINIC
14	96O0JS7E	10/16/2013	CLINIC003	$280.40	CLINIC
15	96O0JS7E	04/08/2013	OPSURGERY036	$7,243.60	OPSURGERY
16	QMPH9WU3	11/04/2013	OPSURGERY041	$1,428.60	OPSURGERY
17	QMPH9WU3	07/03/2013	OPRADIOLOGY070	$3,180.20	OPRADIOLOGY
18	AD0ONT9C	03/21/2013	CLINIC012	$62.00	CLINIC
19	ZLEYIOE7	01/18/2013	PHYSTHERAPY074	$209.40	PHYSTHERAPY
20	ZLEYIOE7	08/20/2013	CLINIC002	$105.20	CLINIC

The DATA step iterates once. It starts by defining hash object PROVIDERLIST. It ends by outputting the contents of hash object PROVIDERLIST. The ORDER: "YES" argument tag on the DECLARE statement specifies that SAS retrieve entries from PROVIDERLIST in ascending order by the values of the key item PROVIDER_ID.

Note that the DATA step does not call the DEFINEDATA method and define data items in PROVIDERLIST. With no data items specified, SAS considers all key items to also be data items. Therefore, the OUTPUT method writes key item PROVIDER_ID to data set CLAIMPROVIDERS. If you did define any data items, you would also have to add PROVIDER_ID to the list of data items. When you have data items and you also want the key item in the output data set, you must add the key item to the list of data items.

Since PROVIDERLIST was not declared to allow multiple sets of data items per key value, the hash object contains only unique values of PROVIDER_ID.

Example 4.7 Finding the Unique Values of a Variable in a Data Set

```
data _null_;
  attrib provider_id length=$15 label='Provider ID';

  declare hash providerlist(dataset: 'claims2013' ordered: 'yes');
  providerlist.definekey('provider_id');
  providerlist.definedone();

  call missing(provider_id);

  rc=providerlist.output(dataset: 'claimproviders');
run;
```

Output 4.13 lists the first 15 observations in data set CLAIMPROVIDERS that the OUTPUT method creates. Example 4.7 found the 75 unique providers in the data set of several thousand observations.

Output 4.13 PROC PRINT of CLAIMPROVIDERS (first 15 observations)

Obs	provider_id
1	CLINIC001
2	CLINIC002
3	CLINIC003
4	CLINIC004
5	CLINIC005
6	CLINIC006
7	CLINIC007
8	CLINIC008
9	CLINIC009
10	CLINIC010
11	CLINIC011
12	CLINIC012
13	CLINIC013
14	CLINIC014
15	CLINIC015

The MULTIDATA: "YES" option can be specified on the DECLARE statement to allow multiple sets of data items per key value. If you specified MULTIDATA: "YES" in Example 4.7, output data set CLAIMPROVIDERS would have the same number of observations as input data set CLAIMS2013, and you would not find the unique values of PROVIDER_ID. For examples that work with multiple sets of data items per key value, see Chapter 5.

Application Example: Ordering Observations by Variables Not Saved to a Data Set Created from a Hash Object

Example 4.8 shows that you can arrange observations in a data set by variables that you do not keep in the data set. It illustrates that the OUTPUT method does not save the variables you specify with the DEFINEKEY method unless you also specify the items in the DEFINEDATA method. It also demonstrates that the OUTPUT method can create a data set whose observations are arranged by the values of the items in the DEFINEKEY method even though you do not specify these items on the DEFINEDATA method.

The goal of Example 4.8 is to select observations from data set EMPLOYEES where the employee has worked for 10 years or more as of December 31, 2012, and to arrange the selected observations alphabetically by last name. However, in the report the employee names should be listed as one variable with first name first and last name last.

Input data set EMPLOYEES contains employment information for a group of employees. Output 4.14 lists the first 20 observations in EMPLOYEES.

Output 4.14 PROC PRINT of EMPLOYEES (first 20 observations)

Obs	empid	empln	empfn	empmi	gender	startdate	emppaylevel
1	6XBIFI	Ramirez	Danielle	N	F	04/21/1989	AIb
2	AWIUME	Thompson	Catherine	D	F	06/18/1986	PIIIa
3	06KH8Q	Chang	William	T	M	07/23/2002	PIIa
4	WA4D7N	Garcia	Breanna	X	F	08/20/1982	AIb
5	OOQT3Z	Jones	Brooke	E	F	08/28/1994	MIIa
6	1JU28B	Smith	Matthew	I	M	08/22/1982	TIIIb
7	V8OARE	Hall	Samuel	B	M	05/25/2010	PIb
8	1GTXQ2	Parker	Nathaniel	S	M	08/12/1996	PIc
9	VPA9EF	Baker	Cheyenne	C	F	02/24/1990	AIIa
10	0IP7L6	Hughes	Alexander	N	M	08/08/1991	TIIb
11	Q1A4SU	Sanchez	Nathaniel	W	M	08/13/1998	TIId
12	ANWFGX	Green	Tyler	I	M	12/04/1991	TIc
13	L1I8Y7	Edwards	Angelica	O	F	11/18/1991	MIIIa
14	TZ6OUB	White	Heather	T	F	01/27/1999	AIIIa
15	235TWE	King	Briana	M	F	10/08/1992	TIIc
16	XYOJC7	Scott	Mark	T	M	01/15/2002	TIIa
17	8TN7WL	Miller	Tyler	J	M	08/31/1998	AIIIc
18	US3DZP	Brown	Sarah	U	F	12/11/2000	TIb
19	ODBAIZ	Jones	Rachel	T	F	10/11/1999	PIIb
20	A4GJG4	Johnson	Angelica	Z	F	01/01/1994	AId

The DATA step in Example 4.8 defines hash object YEARS. Three of the four key items—EMPLN, EMPFN, and EMPMI—defined for YEARS are not defined as data items. The CATX function concatenates these three items to form new variable EMPNAME. The DATA step defines only YEARSCAT as both a key item and a data item.

The SET statement reads data set EMPLOYEES. The number of years of employment is calculated and categorized in variable YEARSCAT. Only observations with EMPYEARS greater than or equal to 10 are selected. The ADD method adds data for the selected observations to hash object YEARS.

The END= option on the SET statement is set to variable EOF. When EOF=1, the DATA step has no more observations to read from EMPLOYEES, and the OUTPUT method creates EMPNAMES10PLUS from hash object YEARS. Because hash object YEARS was defined with the ORDERED: 'YES' argument tag, the observations in EMPNAMES10PLUS are arranged in ascending order by the values of the four key variables, YEARSCAT, EMPLN, EMPFN, and EMPMI. The values of EMPNAME do not affect the order of retrieval from YEARS.

The DATA step saves in data set EMPNAMES10PLUS only the variables listed in the DEFINEDATA method.

Example 4.8 Ordering Observations by Variables Not Saved to a Data Set Created from a Hash Object

```
data _null_;
  attrib empid length=$6
         empname length=$60
         startdate length=8 format=mmddyy10.
         empyears length=8 label='Number of years worked'
         yearscat length=$10 label='Years worked category';

  if _n_=1 then do;
    declare hash years(ordered: 'yes');
    years.definekey('yearscat','empln','empfn','empmi');
    years.definedata('empid','empname','yearscat');
    years.definedone();
    call missing(empid,empname,startdate,empyears,yearscat);
  end;

  set employees end=eof;

  empyears=ceil(('31dec2012'd-startdate)/365.25);

  if empyears ge 10;

  if 10 le empyears lt 15 then yearscat='10+ years';
  else if 15 le empyears lt 20 then yearscat='15+ years';
  else if 20 le empyears lt 25 then yearscat='20+ years';
  else if empyears ge 25 then yearscat='25+ years';
```

```
    empname=catx(' ',empfn,empmi,empln);

    rc=years.add();

    if eof then rc=years.output(dataset: 'empnames10plus');
  run;
```

Output 4.15 lists the first 20 observations in data set EMPNAMES10PLUS. SAS arranged the observations in order first by YEARSCAT and within those values by the elements of the employee's name in this order: EMPLN, EMPFN, and EMPMI.

Output 4.15 PROC PRINT of EMPNAMES10PLUS (first 20 observations)

Obs	empid	empname	yearscat
1	5F9U8L	Alexander X Adams	10+ years
2	YY5RXY	Christopher Z Adams	10+ years
3	Y00PWK	Joshua I Adams	10+ years
4	4A25WB	Lucas M Adams	10+ years
5	KNMW87	Michael G Adams	10+ years
6	P6L8NB	Nathaniel H Adams	10+ years
7	JSSKV0	Nathaniel N Adams	10+ years
8	ACU122	Thomas C Adams	10+ years
9	Y4RIRR	Tiffany W Adams	10+ years
10	FIQ6AU	Jenna I Alexander	10+ years
11	N4NTP7	Lucas W Alexander	10+ years
12	TFYQP8	Samantha W Alexander	10+ years
13	W50SR6	Sean E Alexander	10+ years
14	OAFG6U	Tiffany H Alexander	10+ years
15	FND3UV	Austin H Allen	10+ years
16	ZHFN62	Samantha J Allen	10+ years
17	GKP04R	Alex V Anderson	10+ years
18	X1DSGG	Allison O Anderson	10+ years
19	CUGR7L	Devin Z Anderson	10+ years
20	2WFA3D	Jeremy Y Anderson	10+ years

Application Example: Using Hash Objects to Apply Transactions to Master Data Sets—Part 1

As described in the "Adding, Modifying, and Removing Data from a Hash Object" section above, methods exist to add and remove data from a hash object. An application where you can apply this technique is when you perform the common data processing task of adding and deleting observations from a master data set based on transactions in another data set. Even though it may be more efficient to use techniques other than hash objects to apply transactions to a master data set, Examples 4.9 and 4.10 demonstrate how you can perform this task with a hash object and they illustrate the methods that add and remove data from a hash object.

DATA step programming techniques and statements such as the MODIFY statement, the UPDATE statement, and match-merging by variables that link the master data set and the transaction data set can apply transactions to master data sets. The MODIFY statement modifies your data set in place and does not rewrite it. When you use the UPDATE statement, match-merging, or a hash object, SAS rewrites your data set. Therefore, if your data set is very large and you do not want to rewrite your master data set, evaluate whether you should use a DATA step and the MODIFY statement. Additionally, the size of your master data set may be too large to load into a hash object in memory.

Replacing data in an entry in a hash object is slightly more complicated than adding and removing entries. Example 4.10 shows how to replace data in a hash object using the same master data set as that used in Example 4.9.

The master data set BOOKLIST contains book publishing information for three books. Each book has several chapters and BOOKLIST records the author, due date, and editor for each chapter in the three books. Data set BOOKSADDDEL contains observations to add or delete from BOOKLIST. The master data set and transaction data set are linked by the values of variables BOOKID and CHAPTER.

Output 4.16 lists the contents of BOOKLIST.

Output 4.16 PROC PRINT of BOOKLIST

Obs	bookid	booktitle	chapter	author	duedate	editor
1	NF0586	Health Data Sets	1	Smith, Rebecca	09/08/2013	Williams, Stephanie
2	NF0586	Health Data Sets	2	Williams, Susan	09/08/2013	Williams, Stephanie
3	NF0586	Health Data Sets	3	Torres, Christopher	09/15/2013	Williams, Stephanie
4	NF0586	Health Data Sets	4	Torres, Christopher	09/15/2013	Williams, Stephanie
5	NF0586	Health Data Sets	5	Powell, George	09/15/2013	Williams, Stephanie
6	NF0586	Health Data Sets	6	Thompson, Tonya	09/15/2013	Williams, Stephanie
7	NF0586	Health Data Sets	7	Allen, Linda	09/15/2013	Williams, Stephanie
8	NF0586	Health Data Sets	8	Johnson, Tammy	09/15/2013	Williams, Stephanie
9	NF0586	Health Data Sets	9	Kelly, Melissa	09/15/2013	Williams, Stephanie
10	NF0586	Health Data Sets	10	Thompson, Tonya	09/15/2013	Williams, Stephanie
11	NF2413	Patient Handouts	1	Jones, Robin	08/03/2013	White, Michelle
12	NF2413	Patient Handouts	2	Sanchez, Brandon	08/11/2013	White, Michelle
13	NF2413	Patient Handouts	3	Jones, Robin	08/03/2013	White, Michelle

Obs	bookid	booktitle	chapter	author	duedate	editor
14	NF2413	Patient Handouts	4	Perez, Joshua	08/03/2013	White, Michelle
15	NF2413	Patient Handouts	5	Williams, Nicholas	08/03/2013	White, Michelle
16	NF2413	Patient Handouts	6	Patterson, Mary	08/18/2013	White, Michelle
17	NF2413	Patient Handouts	7	Torres, Christopher	08/11/2013	White, Michelle
18	NF2413	Patient Handouts	8	Robinson, Bonnie	08/11/2013	White, Michelle
19	NF2413	Patient Handouts	9	Brown, Patricia	08/11/2013	White, Michelle
20	NF8141	Medical Writing	1	Clark, Todd	10/06/2013	Patterson, Daniel
21	NF8141	Medical Writing	2	Barnes, David	10/06/2013	Patterson, Daniel
22	NF8141	Medical Writing	3	Young, Richard	09/22/2013	Patterson, Daniel
23	NF8141	Medical Writing	4	Barnes, David	10/06/2013	Patterson, Daniel
24	NF8141	Medical Writing	5	Anderson, Daniel	09/22/2013	Patterson, Daniel
25	NF8141	Medical Writing	6	Anderson, Daniel	09/22/2013	Patterson, Daniel
26	NF8141	Medical Writing	7	Morris, Laura	09/22/2013	Patterson, Daniel
27	NF8141	Medical Writing	8	Powell, George	09/22/2013	Patterson, Daniel

Output 4.17 lists the contents of BOOKSADDDEL.

Output 4.17 PROC PRINT of BOOKSADDDEL

Obs	action	bookid	chapter	booktitle	author	duedate	editor
1	DEL	NF0586	2			.	
2	ADD	NF0586	2	Health Data Sets	King, Weston	09/22/2013	Williams, Stephanie
3	ADD	NF2413	10	Patient Handouts	Moses, Winston	08/24/2013	White, Michelle
4	ADD	NF2413	1	Patient Handouts	Johnson, Rosa	08/17/2013	White, Michelle
5	DEL	NF8141	9			.	
6	ADD	NF8411	11	Medical Writing	Powell, George	09/22/2013	Patterson, Daniel

Example 4.9 allows two types of transactions: add or delete. Therefore, variable ACTION in BOOKSADDDEL can have one of two values: "ADD" or "DEL", which indicate whether to add or delete the matching observation in BOOKLIST. The data in BOOKSADDDEL specify that two observations should be deleted and four should be added.

On its first iteration, the DATA step creates two hash objects, BL and IDS, and each hash object contains data from data set BOOKLIST. Hash object BL contains all the data from data set BOOKLIST. Hash object IDS contains one key item, BOOKIDS, and no data items. Hash object IDS contains the set of unique BOOKID values and thus has only three entries.

Example 4.9 updates the contents of BL with the transactions in BOOKSADDDEL. The last action in the DATA step outputs the updated contents of BL to a new version of data set BOOKLIST. BL has two key items, BOOKID and CHAPTER. These two keys link BOOKLIST and

BOOKSADDDEL. The ALL: "YES" argument tag in the DEFINEDATA method loads all variables in BOOKLIST into BL.

Hash object IDS serves as a lookup table so that it can be determined if the current value of BOOKID as read from the transaction data set exists in BOOKLIST. A condition of this task is that the DATA step must not add a new book to BL. It determines that an observation in BOOKLIST is for a new book if the value of BOOKID is not found in IDS.

IF-THEN statements that test the values of ACTION control whether SAS adds or deletes entries from hash object BL:

- When ACTION='DEL', the DATA step applies the REMOVE method. A non-zero return code from applying this method indicates that the DATA step was not able to delete the entry in BL. When this occurs, the DATA step writes an error message to the SAS log.
- When ACTION='ADD', the DATA step applies the CHECK method to see if the observation in BOOKSADDDEL currently being processed by the DATA step has a match by BOOKID and CHAPTER in BL. If a match is found in BL, no update to the hash object is made and the DATA step writes an error message to the SAS log. If a match is not found in BL, the CHECK method looks in IDS to see if the current value of BOOKID read from BOOKSADDDEL exists in BOOKLIST. If it does, the ADD method adds the entry to BL. If it does not, the DATA step writes an error message to the SAS log.

When the END= variable, EOF, equals 1, the DATA step has read the last observation in transaction data set BOOKSADDDEL, and it has made all updates to hash object BL. The DATA step then applies the output method and overwrites existing data set BOOKLIST with an updated version that it derives from hash object BL.

Example 4.9 Using a Hash Object to Apply Transactions to a Master Data Set

```
data _null_;
  attrib bookid  length=$6
         booktitle length=$18
         chapter length=8
         author  length=$20
         duedate length=8 format=mmddyy10.
         editor  length=$20;

  if _n_=1 then do;
    declare hash bl(dataset: 'booklist', ordered: 'y');
    bl.definekey('bookid','chapter');
    bl.definedata(all: 'yes');
    bl.definedone();

    declare hash ids(dataset: 'booklist');
    ids.definekey('bookid');
    ids.definedone;
```

```
    call missing(bookid,booktitle,chapter,author,duedate,editor);
  end;

  set booksadddel end=eof;

  if action='DEL' then do;
    rc=bl.remove();
    if rc ne 0 then
        putlog "ERROR: This obs could not be deleted: " bookid=chapter=;
  end;
  else if action='ADD' then do;
    rc=bl.check();
    if rc=0 then putlog
    "ERROR:This obs already exists so it cannot be added: " bookid=chapter=;
    else do;
      rc=ids.check();
      if rc=0 then rc=bl.add();
      else putlog
"ERROR: BOOKID=" bookid "does not exist in BOOKLIST so it cannot be added";
    end;
  end;

  if eof then rc=bl.output(dataset: 'booklist');
run;
```

The SAS log provides information about the updates made to BOOKLIST.

```
NOTE: There were 27 observations read from the data set WORK.BOOKLIST.
NOTE: There were 27 observations read from the data set WORK.BOOKLIST.
ERROR: This obs already exists so it cannot be added: bookid=NF2413 chapter=1
ERROR: This obs could not be deleted: bookid=NF8141 chapter=9
ERROR: BOOKID=NF8411 does not exist in BOOKLIST so it cannot be added
NOTE: The data set WORK.BOOKLIST has 28 observations and 6 variables.
NOTE: There were 6 observations read from the data set WORK.BOOKSADDDEL.
```

The first two NOTE statements list the number of observations in master data set BOOKLIST. These notes correspond to SAS loading hash objects BL and IDS with data from BOOKLIST. The third NOTE statement lists the number of observations that the DATA step outputs to the updated version of BOOKLIST by the OUTPUT method. These notes appear before the fourth note that states how many observations that the SET statement read from BOOKSADDDEL.

Three ERROR messages show the transactions that the DATA step did not apply. DATA step statements supply the text of these ERROR messages.

Output 4.18 shows the updated version of data set BOOKLIST.

Output 4.18 PROC PRINT of Updated Version of BOOKLIST

Obs	bookid	booktitle	chapter	author	duedate	editor
1	NF0586	Health Data Sets	1	Smith, Rebecca	09/08/2013	Williams, Stephanie
2	NF0586	Health Data Sets	2	King, Weston	09/22/2013	Williams, Stephanie
3	NF0586	Health Data Sets	3	Torres, Christopher	09/15/2013	Williams, Stephanie
4	NF0586	Health Data Sets	4	Torres, Christopher	09/15/2013	Williams, Stephanie
5	NF0586	Health Data Sets	5	Powell, George	09/15/2013	Williams, Stephanie
6	NF0586	Health Data Sets	6	Thompson, Tonya	09/15/2013	Williams, Stephanie
7	NF0586	Health Data Sets	7	Allen, Linda	09/15/2013	Williams, Stephanie
8	NF0586	Health Data Sets	8	Johnson, Tammy	09/15/2013	Williams, Stephanie
9	NF0586	Health Data Sets	9	Kelly, Melissa	09/15/2013	Williams, Stephanie
10	NF0586	Health Data Sets	10	Thompson, Tonya	09/15/2013	Williams, Stephanie
11	NF2413	Patient Handouts	1	Jones, Robin	08/03/2013	White, Michelle
12	NF2413	Patient Handouts	2	Sanchez, Brandon	08/11/2013	White, Michelle
13	NF2413	Patient Handouts	3	Jones, Robin	08/03/2013	White, Michelle
14	NF2413	Patient Handouts	4	Perez, Joshua	08/03/2013	White, Michelle
15	NF2413	Patient Handouts	5	Williams, Nicholas	08/03/2013	White, Michelle
16	NF2413	Patient Handouts	6	Patterson, Mary	08/18/2013	White, Michelle
17	NF2413	Patient Handouts	7	Torres, Christopher	08/11/2013	White, Michelle
18	NF2413	Patient Handouts	8	Robinson, Bonnie	08/11/2013	White, Michelle
19	NF2413	Patient Handouts	9	Brown, Patricia	08/11/2013	White, Michelle
20	NF2413	Patient Handouts	10	Moses, Winston	08/24/2013	White, Michelle
21	NF8141	Medical Writing	1	Clark, Todd	10/06/2013	Patterson, Daniel
22	NF8141	Medical Writing	2	Barnes, David	10/06/2013	Patterson, Daniel
23	NF8141	Medical Writing	3	Young, Richard	09/22/2013	Patterson, Daniel
24	NF8141	Medical Writing	4	Barnes, David	10/06/2013	Patterson, Daniel
25	NF8141	Medical Writing	5	Anderson, Daniel	09/22/2013	Patterson, Daniel
26	NF8141	Medical Writing	6	Anderson, Daniel	09/22/2013	Patterson, Daniel
27	NF8141	Medical Writing	7	Morris, Laura	09/22/2013	Patterson, Daniel
28	NF8141	Medical Writing	8	Powell, George	09/22/2013	Patterson, Daniel

Application Example: Using Hash Objects to Apply Transactions to Master Data Sets—Part 2

The code in Example 4.10 continues the theme of Example 4.9 by performing the common data processing task of applying transactions to a master data set. In addition to applying the ADD and

REMOVE methods as Example 4.9 presents, Example 4.10 applies the REPLACE method to replace data items in a hash object.

The master data set BOOKLIST, which is the same data set that Example 4.9 used as input, and was shown in Output 4.16, contains book publishing information on three books. Each book has several chapters. BOOKLIST records the author, due date, and editor for each chapter. The transactions data set is BOOKEDITS. The master and transaction data sets are linked by the values of variables BOOKID and CHAPTER.

Variable ACTION in BOOKEDITS can have three values: “ADD”, “DEL”, or “MOD”. These values code the type of transaction to apply to data set BOOKLIST: add an entry to the hash object; remove an entry from the hash object; or modify an entry in the hash object.

The first four observations in BOOKEDITS specify changes to make to existing observations in BOOKLIST. The remaining six observations in BOOKEDITS specify that two observations should be deleted and four should be added. The additions and deletions are identical to those specified in Example 4.9.

Output 4.19 lists the observations in BOOKEDITS.

Output 4.19 PROC PRINT of BOOKEDITS

Obs	action	bookid	chapter	newbooktitle	newauthor	newduedate	neweditor
1	MOD	NF0586	8		Lester, Jose	09/22/2013	
2	MOD	NF0586	12			.	White, Michelle
3	MOD	NF4213	8			09/01/2013	
4	MOD	NF2413	.			.	
5	DEL	NF0586	2			.	
6	ADD	NF0586	2	Health Data Sets	King, Weston	09/22/2013	Williams, Stephanie
7	ADD	NF2413	10	Patient Handouts	Moses, Winston	08/24/2013	White, Michelle
8	ADD	NF2413	1	Patient Handouts	Johnson, Rosa	08/17/2013	White, Michelle
9	DEL	NF8141	9			.	
10	ADD	NF8411	11	Medical Writing	Powell, George	09/22/2013	Patterson, Daniel

On its first iteration, the DATA step loads two hash objects, BL and IDS, with data from data set BOOKLIST. Hash object BL has two keys, BOOKID and CHAPTER, and four data items: BOOKTITLE, AUTHOR, DUEDATE, and EDITOR. The DATA step updates the contents of BL with the transactions in BOOKEDITS.

Hash object IDS contains one key item and no data items. The key item is BOOKID. IDS serves as a lookup table so that it can be determined if the value of BOOKID in the transaction data set exists in BOOKLIST. The DATA step must not add a new book to BL. It concludes that an observation in BOOKLIST is for a new book if it does not find the value of BOOKID in IDS.

After the DATA step has read all observations in BOOKEDITS, it applies the OUTPUT method to overwrite existing data set BOOKLIST with an updated version that it derives from hash object BL.

Example 4.10 lists each of the data items in the DEFINEDATA method. Compare that to Example 4.9 where the DEFINEDATA argument tag ALL: "YES" loaded all variables in BOOKLIST as data in BL. Although the ALL: "YES" argument tag could be used in this example, it is not used because of the code in the statement with the REPLACE method that manages the modification of data items in BL.

When you use the ALL: "YES" argument tag in the DEFINEDATA method, the order of the data items is the same as the order of the variables in the input data set. When you specify the DATA: argument tag values in the REPLACE method, you have to specify your values in this same order. The order of the variables may not be readily available or may change if your data set is rewritten. If you have a lot of variables in your data set, it might be simpler to explicitly list the items in the DEFINEKEY and DEFINEDATA methods so that you do not have to match the order of the variables in the PDV. Instead you match the order of the items listed in the DEFINEDATA method call.

The transaction data set in this example, BOOKEDITS, has different variable names than the variable names in BOOKLIST, as shown in Output 4.20. The variable names were changed because of the default actions of the FIND method. When the FIND method executes, it copies the values from the hash object into variables with the same names as the data items. With the names different, the values from the two sources can be compared.

When you write code to update a data set, you must consider how to handle missing values in the transaction data set. You might or might not want your code to replace a nonmissing value in your master data set with a missing value that is present in your transaction data set. The code in Example 4.10 does not replace a nonmissing value in the corresponding data item in BL when a missing value for a variable in the transaction data set exists.

The COALESCE and COALESCEC functions used in the REPLACE method ensure that missing values in the transaction data set do not replace nonmissing values in the corresponding data items in BL. These functions return the first nonmissing value in the series of arguments supplied to them. The first argument to all four calls to these functions is the variable from the transaction data set.

As in Example 4.9, IF-THEN statements test the values of ACTION. The values of ACTION control whether entries are added to, deleted from, or modified in hash object BL. The same actions occur in Example 4.10 as they did in Example 4.9 when ACTION='DEL' or ACTION='ADD'. When ACTION='MOD', Example 4.10 modifies data item values based on the results of the COALESCE and COALESCEC function calls.

Example 4.10 Using a Hash Object to Apply Transactions to a Master Data Set

```
data _null_;
  attrib bookid   length=$6
         booktitle length=$18
         chapter  length=8
         author   length=$20
         duedate  length=8 format=mmddyy10.
         editor   length=$20;

  if _n_=1 then do;
    declare hash bl(dataset: 'booklist', ordered: 'y');
    bl.definekey('bookid','chapter');
    bl.definedata('bookid','chapter','booktitle','author','duedate',
                  'editor');
    bl.definedone();

    declare hash ids(dataset: 'booklist');
    ids.definekey('bookid');
    ids.definedone();

    call missing(bookid,booktitle,chapter,author,duedate,editor);
  end;

  set bookedits end=eof;

  if action='MOD' then do;
    rc=bl.find();
    if rc ne 0 then putlog
   "ERROR: This obs does not exist so it cannot be edited:" bookid= chapter=;
    else rc=bl.replace(key: bookid, key: chapter,
                       data: bookid, data: chapter,
                       data: coalescec(newbooktitle,booktitle),
                       data: coalescec(newauthor,author),
                       data: coalesce(newduedate,duedate),
                       data: coalescec(neweditor,editor));
  end;
  else if action='DEL' then do;
    rc=bl.remove();
    if rc ne 0 then
         putlog "ERROR: This obs could not be deleted: " bookid= chapter=;
  end;
  else if action='ADD' then do;
    rc=bl.check();
    if rc=0 then putlog
   "ERROR: This obs already exists so it cannot be added:" bookid=chapter=;
    else do;
      rc=ids.check();
      if rc=0 then do;
        booktitle=newbooktitle;
        author=newauthor;
```

```
         duedate=newduedate;
         editor=neweditor;
         rc=bl.add();
       end;
       else putlog
    "ERROR: BOOKID=" bookid "does not exist in BOOKLIST so it cannot
             be added";
    end;
  end;

  if eof then rc=bl.output(dataset: 'booklist');
run;
```

The SAS log provides information about the updates made to BOOKLIST.

```
NOTE: There were 27 observations read from the data set WORK.BOOKLIST.
NOTE: There were 27 observations read from the data set WORK.BOOKLIST.
ERROR: This obs does not exist so it cannot be edited: bookid=NF0586
chapter=12
ERROR: This obs does not exist so it cannot be edited: bookid=NF4213
chapter=8
ERROR: This obs does not exist so it cannot be edited: bookid=NF2413
chapter=.
ERROR: This obs already exists so it cannot be added: bookid=NF2413
chapter=1
ERROR: This obs could not be deleted: bookid=NF8141 chapter=9
ERROR: BOOKID=NF8411 does not exist in BOOKLIST so it cannot be added
NOTE: The data set WORK.BOOKLIST has 28 observations and 4 variables.
NOTE: There were 10 observations read from the data set WORK.BOOKEDITS.
```

The first two NOTE statements list the number of observations in master data set BOOKLIST. These notes correspond to SAS loading hash objects BL and IDS with data from BOOKLIST. The third NOTE statement lists the number of observations the DATA step outputs to the updated version of BOOKLIST. These three notes appear before the fourth note that states how many observations the SET statement read from BOOKEDITS.

Six ERROR messages generated by the code in the DATA step show the transactions that the DATA step could not apply. The first three result from ACTION='MOD' observations. The last three are the same as the error messages that Example 4.9 generated.

Output 4.20 lists the updated version of BOOKLIST.

Output 4.20 PROC PRINT of Updated Version of Data Set BOOKLIST

Obs	bookid	chapter	booktitle	author	duedate	editor
1	NF0586	1	Health Data Sets	Smith, Rebecca	09/08/2013	Williams, Stephanie
2	NF0586	2	Health Data Sets	King, Weston	09/22/2013	Williams, Stephanie
3	NF0586	3	Health Data Sets	Torres, Christopher	09/15/2013	Williams, Stephanie
4	NF0586	4	Health Data Sets	Torres, Christopher	09/15/2013	Williams, Stephanie
5	NF0586	5	Health Data Sets	Powell, George	09/15/2013	Williams, Stephanie
6	NF0586	6	Health Data Sets	Thompson, Tonya	09/15/2013	Williams, Stephanie
7	NF0586	7	Health Data Sets	Allen, Linda	09/15/2013	Williams, Stephanie
8	NF0586	8	Health Data Sets	Lester, Jose	09/22/2013	Williams, Stephanie
9	NF0586	9	Health Data Sets	Kelly, Melissa	09/15/2013	Williams, Stephanie
10	NF0586	10	Health Data Sets	Thompson, Tonya	09/15/2013	Williams, Stephanie
11	NF2413	1	Patient Handouts	Jones, Robin	08/03/2013	White, Michelle
12	NF2413	2	Patient Handouts	Sanchez, Brandon	08/11/2013	White, Michelle
13	NF2413	3	Patient Handouts	Jones, Robin	08/03/2013	White, Michelle
14	NF2413	4	Patient Handouts	Perez, Joshua	08/03/2013	White, Michelle
15	NF2413	5	Patient Handouts	Williams, Nicholas	08/03/2013	White, Michelle
16	NF2413	6	Patient Handouts	Patterson, Mary	08/18/2013	White, Michelle
17	NF2413	7	Patient Handouts	Torres, Christopher	08/11/2013	White, Michelle
18	NF2413	8	Patient Handouts	Robinson, Bonnie	08/11/2013	White, Michelle
19	NF2413	9	Patient Handouts	Brown, Patricia	08/11/2013	White, Michelle
20	NF2413	10	Patient Handouts	Moses, Winston	08/24/2013	White, Michelle
21	NF8141	1	Medical Writing	Clark, Todd	10/06/2013	Patterson, Daniel
22	NF8141	2	Medical Writing	Barnes, David	10/06/2013	Patterson, Daniel
23	NF8141	3	Medical Writing	Young, Richard	09/22/2013	Patterson, Daniel
24	NF8141	4	Medical Writing	Barnes, David	10/06/2013	Patterson, Daniel
25	NF8141	5	Medical Writing	Anderson, Daniel	09/22/2013	Patterson, Daniel
26	NF8141	6	Medical Writing	Anderson, Daniel	09/22/2013	Patterson, Daniel
27	NF8141	7	Medical Writing	Morris, Laura	09/22/2013	Patterson, Daniel
28	NF8141	8	Medical Writing	Powell, George	09/22/2013	Patterson, Daniel

Application Example: Summarizing Data with the Hash Iterator Object

Example 4.11 shows how you can summarize information about key items in your hash object. The SUMINC argument tag on the DECLARE statement that defines the hash object tells SAS that you want to tally summaries for each key value. When you use the SUMINC argument tag, you usually associate a hash object with a hash iterator object. As the hash iterator object traverses the hash object, the DATA step tallies summary data for each key value it processes. The name of a DATA step variable is supplied to the SUMINC argument. The SUMINC variable's value is added to the summary for each key value each time a FIND, CHECK, or REF method call accesses the key value.

While the SUMINC feature is limited in its usage compared to the features that SAS language statements and procedures like MEANS and FREQ can provide, it can be useful when you need to quickly summarize data or tally the number of key values in a data set. The DECLARE statement allows only one SUMINC: argument tag.

Example 4.11 reads the CLAIMS2013 data set that Example 4.7 also processes. Data set CLAIMS2013 contains medical claims data. Each observation contains data for one medical claim for one patient on one date. Variable PROVIDER_ID contains the ID for the facility that provided care to the patient. The nonnumeric part of each value of PROVIDER_ID contains text that identifies five types of providers: Clinic (CLINIC), Lab (LAB), Outpatient Radiology (OPRADIOLOGY), Outpatient Surgery (OPSURGERY), and Physical Therapy (PHYSTHERAPY). These values are saved in variable PROVIDER_TYPE. The numeric part of the PROVIDER_ID value is an ID assigned that uniquely identifies the facility within that provider type. Variable CHARGE contains the charge applied to the patient's visit.

The goal of Example 4.11 is to sum the values of CHARGE for each provider and for each type of provider. The DATA step saves the provider summaries in data set PROVIDERS. It saves the summaries for the type of provider in data set TYPES.

Output 4.21 lists the first 20 observations in CLAIMS2013.

Output 4.21 PROC PRINT of CLAIMS2013 (first 20 observations)

Obs	ptid	claimdate	provider_id	charge	provider_type
1	4DGNWU3Z	04/12/2013	OPSURGERY039	$7,902.00	OPSURGERY
2	4DGNWU3Z	02/07/2013	PHYSTHERAPY071	$330.40	PHYSTHERAPY
3	XZRQTXGU	04/17/2013	OPRADIOLOGY069	$1,517.60	OPRADIOLOGY
4	JT1Z17GE	08/22/2013	CLINIC016	$53.00	CLINIC
5	THSALN2J	09/25/2013	OPRADIOLOGY070	$450.20	OPRADIOLOGY
6	THSALN2J	11/11/2013	CLINIC014	$229.80	CLINIC
7	THSALN2J	08/15/2013	CLINIC010	$102.20	CLINIC
8	THSALN2J	09/09/2013	LAB049	$147.40	LAB
9	DWJSEGFQ	07/31/2013	PHYSTHERAPY074	$381.40	PHYSTHERAPY
10	DWJSEGFQ	10/04/2013	CLINIC023	$65.80	CLINIC
11	L2PAT59T	10/16/2013	OPSURGERY044	$3,373.80	OPSURGERY
12	GE356WUG	10/07/2013	OPSURGERY035	$7,128.20	OPSURGERY
13	GE356WUG	07/24/2013	CLINIC005	$104.60	CLINIC

Obs	ptid	claimdate	provider_id	charge	provider_type
14	96O0JS7E	10/16/2013	CLINIC003	$280.40	CLINIC
15	96O0JS7E	04/08/2013	OPSURGERY036	$7,243.60	OPSURGERY
16	QMPH9WU3	11/04/2013	OPSURGERY041	$1,428.60	OPSURGERY
17	QMPH9WU3	07/03/2013	OPRADIOLOGY070	$3,180.20	OPRADIOLOGY
18	AD0ONT9C	03/21/2013	CLINIC012	$62.00	CLINIC
19	ZLEYIOE7	01/18/2013	PHYSTHERAPY074	$209.40	PHYSTHERAPY
20	ZLEYIOE7	08/20/2013	CLINIC002	$105.20	CLINIC

Example 4.11 defines one hash object and one hash iterator object for each of the two summaries. Both hash objects, HTYPES and HPROVIDERS, define CHARGE as the SUMINC variable. This means that each time a FIND, CHECK, or REF method executes for a key value, the DATA step adds the associated value of data item CHARGE to the summary for the key value.

The DATA step iterates once. It starts by defining the two hash objects and two hash iterator objects. The DO UNTIL loop reads the observations in CLAIMS2013. The REF method executes twice on each iteration of the DO UNTIL loop, once for each of the two hash objects.

When SAS encounters a key value for the first time, the REF method adds the key value to the hash object and initializes the summary for the key value with the current value of CHARGE. For subsequent occurrences of the key value, the REF method does not add any more instances of the key value to the hash object, but SAS does add the current value of CHARGE to the summary for the key value.

By default, all of the key values in a hash object are unique. That means that only one set of data items is allowed per key value. The values of the key items in both hash objects in Example 4.11 are not unique in data set CLAIMS2013. You could add the MULTIDATA: "YES" argument tag to the DECLARE statement for each hash object, which would allow multiple sets of data items per key value in each hash object. However, in this example, the MULTIDATA: "YES" specification would result in a hash object with the same number of key values as the number of observations in the CLAIMS2013 data set. You would not end up with a hash object that collected only the unique key values and the sum of the SUMINC variable values for each unique key value.

When the DO UNTIL loop finishes, hash object HTYPES has five entries, one for each of the five provider types. It also contains the sum of CHARGE for each of the five provider types. Similarly, hash object HPROVIDERS has 75 entries, one for each of the facilities, and it contains the sum of CHARGE for each of the 75 facilities.

The two hash iterator objects traverse the two hash objects in the two DO WHILE loops. With the ORDERED: "A" argument tag specified, SAS retrieves data from the hash objects in ascending key value order.

In each DO WHILE loop, SAS executes the SUM method and an OUTPUT statement for each entry in the hash object. The SUM method retrieves the summary for the key value currently being processed and assigns this value to the DATA step variable the method specifies on its SUM argument tag. Both DO WHILE loops assign the summary values to variable TOTAL. The OUTPUT statements write the observation to the appropriate data set.

Example 4.11 uses OUTPUT statements instead of calls to the OUTPUT method because the only way to retrieve summary values from the hash object is with the SUM method. If you replaced a DO WHILE loop with an OUTPUT method call, your output data set would contain only the key variable; the data set created from hash object HTYPES would contain only variable PROVIDER_TYPE and the data set created from hash object HPROVIDERS would contain only variable PROVIDER_ID.

Example 4.11 Summarizing Data with the Hash Iterator Object

```
data types(keep=provider_type total)
     providers(keep=provider_id total);

  attrib provider_id length=$15 label='Provider ID'
         provider_type length=$11 label='Provider Type'
         total length=8 format=dollar16.2;

  declare hash htypes(suminc: 'charge', ordered: 'a');
  declare hiter itertypes('htypes');
  htypes.definekey('provider_type');
  htypes.definedone();

  declare hash hproviders(suminc: 'charge', ordered: 'a');
  declare hiter iterprov('hproviders');
  hproviders.definekey('provider_id');
  hproviders.definedone();

  do until(eof);
    set claims2013 end=eof;
    htypes.ref();
    hproviders.ref();
  end;

  rc = itertypes.first();
  do while (rc = 0);
    htypes.sum(sum: total);
    output types;
    rc = itertypes.next();
  end;
```

```
    rc = iterprov.first();
    do while (rc = 0);
      hproviders.sum(sum: total);
      output providers;
      rc = iterprov.next();
    end;
  run;
```

Output 4.22 shows the contents of data set TYPES.

Output 4.22 PROC PRINT of TYPES

Obs	provider_type	total
1	CLINIC	$12,491,639.40
2	LAB	$9,636,894.20
3	OPRADIOLOGY	$39,431,305.40
4	OPSURGERY	$176,611,772.00
5	PHYSTHERAPY	$3,566,354.40

Output 4.23 shows the first 10 observations of data set PROVIDERS.

Output 4.23 PROC PRINT of PROVIDERS (first 10 observations)

Obs	provider_id	total
1	CLINIC001	$360,983.60
2	CLINIC002	$367,128.20
3	CLINIC003	$360,311.20
4	CLINIC004	$381,567.20
5	CLINIC005	$369,150.60
6	CLINIC006	$364,069.80
7	CLINIC007	$375,943.00
8	CLINIC008	$383,226.60
9	CLINIC009	$356,274.40
10	CLINIC010	$371,226.60

Application Example: Summarizing Hierarchically Related Data

The two input data sets in Example 4.12 are hierarchically related. The goal is to process the data in a manner that is similar to BY-group processing. Each observation in the higher level data set has one or more detail observations in the lower level data set. The DATA step summarizes the detail observations for each group and then adds the summaries to each of the detail observations.

The data sets used in this example are from a survey that gathered data about households and the people living in them. Data set HH is the higher level data set, and each observation in HH stores general information about one household. Data set PERSONS stores data about each person who lives in each household present in HH. Variable HHID, which is common to both data sets, uniquely identifies each household. Each household can have one or more persons living in the household. Two variables uniquely identify the observations in PERSONS: the household ID variable HHID and the person ID variable PERSONID.

The persons interviewed are assigned a sequential ID value starting with 1. All households have at least one household member and an observation where PERSONID='P01'.

Example 3.13 and Example 5.14 also use data sets HH and PERSONS.

The DATA step in Example 4.12 determines four statistics:

- the total number of persons in a household
- the highest level of education in a household
- the total household income
- the age of the person with PERSONID='P01' in each household

The goal is to attach to each person's observations the summary information for the household in which the person lives. Variable HHTYPE extracted from HH is also added to each person's observation. Therefore, the output data set contains observations at the person level.

Output 4.24 shows the contents of HH. Data set HH contains data for 10 households.

Output 4.24 PROC PRINT of HH

Obs	hhid	tract	surveydate	hhtype
1	HH01	CS	07/09/2012	Owner
2	HH02	CN	03/11/2012	Renter
3	HH03	CS	05/20/2012	Owner
4	HH04	SW	01/12/2012	Owner
5	HH05	NE	10/17/2012	Renter
6	HH06	NE	05/15/2012	Owner
7	HH07	SW	02/02/2012	Owner
8	HH08	NE	04/09/2012	Renter
9	HH09	CE	11/01/2012	Owner
10	HH10	CN	03/31/2012	Owner

Data set PERSONS contains at least one observation for every household in HH. Output 4.25 shows the contents of PERSONS.

Output 4.25 PROC PRINT of PERSONS

Obs	hhid	personid	age	gender	income	educlevel
1	HH01	P01	68	M	$52,000	12
2	HH01	P02	68	F	$23,000	12
3	HH02	P01	42	M	$168,100	22
4	HH03	P01	79	F	$38,000	10
5	HH04	P01	32	F	$56,000	16
6	HH04	P02	31	M	$72,000	18
7	HH04	P03	5	F	$0	0
8	HH04	P04	2	F	$0	0
9	HH05	P01	26	M	$89,000	22
10	HH06	P01	56	M	$123,000	18
11	HH06	P02	48	F	$139,300	18
12	HH06	P03	17	F	$5,000	11
13	HH07	P01	48	M	$90,120	16
14	HH07	P02	50	F	$78,000	18
15	HH08	P01	59	F	$55,500	16
16	HH09	P01	32	F	$48,900	14
17	HH09	P02	10	M	$0	5
18	HH10	P01	47	F	$78,000	16
19	HH10	P02	22	F	$32,000	16
20	HH10	P03	19	M	$20,000	12
21	HH10	P04	14	M	$0	8
22	HH10	P05	12	F	$0	6
23	HH10	P06	9	F	$0	4

On its first iteration, the DATA step defines hash object P and hash iterator object PI. It associates PI with P, and loads all observations from PERSONS into P. Variable HHID is renamed to PHHID so that the HHID value from PERSONS can be compared to the value in HHID without replacing the value found in HHID when the DATA step calls the SETCUR and NEXT methods. The entries in hash object P are keyed and ordered by PHHID and PERSONID.

Also on its first iteration, the DATA step creates hash object HHSUMM. This hash object's purpose is as a place to store the summary information for each household. The DATA step does not load data from a data set into HHSUMM. Instead, once a household's information is summarized, the ADD method writes the summaries to HHSUMM. Hash object HHSUMM is keyed by HHID.

The last part of the IF-THEN block that executes on the DATA step's first iteration processes two DO UNTIL loops. The outer DO UNTIL loop reads data set HH with the SET statement. The inner

DO UNTIL loop retrieves each entry in P for the household currently being processed, and it summarizes each household's person-level data.

Since the entries in P are ordered by PHHID and PERSONID, retrieval by the hash iterator object with the NEXT method is in order by the values of PERSONID within the household. The DO UNTIL loop stops when the values of HHID and PHHID are unequal or when the return code from the NEXT method is not zero, which occurs after SAS reads the last entry in P.

The values of HHID and PHHID are unequal when the current entry read from P belongs to the next household, which happens when SAS has read all of the person entries for the household. This condition causes the DO UNTIL loop to end. The ADD method then writes the summarized information for the household, and writes the value of HHTYPE for the household to HHSUMM. Control returns to the top of the outer DO UNTIL loop, and SAS reads the observation for the next household from data set HH. The SETCUR method retrieves data for the first person in this new household. The inner DO UNTIL loop processes the person-level data for the new household.

Each iteration of the outer DO UNTIL loop initializes the summary variables to 0. Variable NPERSONS tallies the total number of people in the household. Variable HHINCOME calculates the total income earned by all members of the household. Variable HIGHESTED is the maximum value of EDUCLEVEL in the household. Variable AGEP01 is the value of AGE for the observation where PERSONID='P01'.

Two executable statements follow the IF-THEN block. The SET statement reads each observation from data set PERSONS. The FIND method retrieves from HHSUMM the household summary information to which the person belongs.

Example 4.12 Summarizing Hierarchically Related Data

```
data personsummary;
  attrib hhid length=$4
         phhid length=$4
         hhtype length=$10
         phhid length=$4
         personid length=$4
         age length=3
         gender length=$1
         income length=8 format=dollar12.
         educlevel length=3
         npersons length=3
         highested length=3
         hhincome  length=8 format=dollar12.
         agep01 length=3;

  if _n_=1 then do;
    declare hash p(dataset: 'persons(rename=(hhid=phhid))',ordered: 'yes');
    declare hiter pi('p');
    p.definekey('phhid','personid');
```

```
      p.definedata('phhid','personid','age','gender','income',
                   'educlevel');
      p.definedone();
      declare hash hhsumm(ordered: 'yes');
      hhsumm.definekey('hhid');
      hhsumm.definedata('hhid','hhtype','npersons','highested',
                        'hhincome',
                        'agep01');
      hhsumm.definedone();

      call missing(phhid,age,gender,educlevel,npersons,highested,
                   hhincome,agep01);

      do until(eof);
        set hh(keep=hhid hhtype) end=eof;
        npersons=0;
        hhincome=0;
        highested=0;

        rc=pi.setcur(key: hhid, key: 'P01');
        do until(rc ne 0 or hhid ne phhid);
          if rc=0 then do;
            if personid='P01' then agep01=age;
            npersons+1;
            hhincome+income;
            if educlevel gt highested then highested=educlevel;
          end;
          rc=pi.next();
        end;
        rc=hhsumm.add();
      end;
    end;

    keep hhid hhtype personid age gender educlevel npersons highested
         hhincome agep01;

    set persons;
    rc=hhsumm.find();
  run;
```

Output data set PERSONSUMMARY contains the same number of observations as PERSONS. Output 4.26 shows the contents of PERSONSUMMARY. Attached to each person's observation are summaries for the household in which the person is a member.

Output 4.26 PROC PRINT of PERSONSUMMARY

Obs	hhid	hhtype	personid	age	gender	educlevel	npersons	highested	hhincome	agep01
1	HH01	Owner	P01	68	M	12	2	12	$75,000	68
2	HH01	Owner	P02	68	F	12	2	12	$75,000	68
3	HH02	Renter	P01	42	M	22	1	22	$168,100	42
4	HH03	Owner	P01	79	F	10	1	10	$38,000	79
5	HH04	Owner	P01	32	F	16	4	18	$128,000	32
6	HH04	Owner	P02	31	M	18	4	18	$128,000	32
7	HH04	Owner	P03	5	F	0	4	18	$128,000	32
8	HH04	Owner	P04	2	F	0	4	18	$128,000	32
9	HH05	Renter	P01	26	M	22	1	22	$89,000	26
10	HH06	Owner	P01	56	M	18	3	18	$267,300	56
11	HH06	Owner	P02	48	F	18	3	18	$267,300	56
12	HH06	Owner	P03	17	F	11	3	18	$267,300	56
13	HH07	Owner	P01	48	M	16	2	18	$168,120	48
14	HH07	Owner	P02	50	F	18	2	18	$168,120	48
15	HH08	Renter	P01	59	F	16	1	16	$55,500	59
16	HH09	Owner	P01	32	F	14	2	14	$48,900	32
17	HH09	Owner	P02	10	M	5	2	14	$48,900	32
18	HH10	Owner	P01	47	F	16	6	16	$130,000	47
19	HH10	Owner	P02	22	F	16	6	16	$130,000	47
20	HH10	Owner	P03	19	M	12	6	16	$130,000	47
21	HH10	Owner	P04	14	M	8	6	16	$130,000	47
22	HH10	Owner	P05	12	F	6	6	16	$130,000	47
23	HH10	Owner	P06	9	F	4	6	16	$130,000	47

Chapter 5: Hash Objects with Multiple Sets of Data Items per Key Value

The default definition of a hash object allows only one set of data items per key value. The examples in the previous chapters were defined in this manner. However, situations exist where it is necessary to have or to look for multiple sets of data items per key value. SAS has argument tags and methods specific for these types of applications. This chapter focuses on how you can define and use hash objects that have multiple sets of data items per key value.

Understanding the Concepts of Duplicate Key Values and Multiple Sets of Data Items per Key Value in a Hash Object

This book repeatedly uses the terminology of multiple sets of data items per key value. To illustrate this concept, consider a data set that contains the schedule for health care professionals staffing a health care event that takes place over a few days. Each observation in the data set contains information about a single shift for one employee. Each employee can work one or more shifts at the event.

Figure 5.1 lists 10 observations from data set OCTOBERVENT. The entire data set contains the schedule information for 52 employees assigned to work at a health screening event on October 22 and 23, 2013. Employees are assigned to a morning ("AM") or afternoon ("PM") shift to perform one of six tasks: cholesterol screening; counseling; dietary advice; immunization; pharmaceutical counseling; and measurement of vital health indicators such as blood pressure and weight.

When the key item is EMPID:

- EMPID 3VGRKR has one set of data items.
- EMPIDs 14ZN75, 1CV01H, and 2S57ZI have two sets of data items.
- EMPID 02QYJG has three sets of data items.

When the key items are EVENTDATE and SHIFT:

- EVENTDATE 10/22/2013 AM shift has four sets of data items.
- EVENTDATE 10/22/2013 PM shift has three sets of data items.
- EVENTDATE 10/23/2013 AM shift has two sets of data items.
- EVENTDATE 10/23/2013 PM shift has one set of data items.

When the key items are EMPID, EVENTDATE, SHIFT, and ACTIVITY, no combination of the key values has multiple sets of data items.

Figure 5.1 Depicting Multiple Sets of Data Items per Key Value

Obs	empid	empname	eventdate	shift	activity
1	02QYJG	Howard, Rachel Y.	10/22/2013	AM	Cholesterol
2	02QYJG	Howard, Rachel Y.	10/22/2013	PM	Cholesterol
3	02QYJG	Howard, Rachel Y.	10/23/2013	AM	Counseling
4	14ZN75	Miller, Sierra Q.	10/22/2013	AM	Counseling
5	14ZN75	Miller, Sierra Q.	10/22/2013	PM	Vitals
6	1CV01H	Brown, Lindsey V.	10/22/2013	AM	Immunization
7	1CV01H	Brown, Lindsey V.	10/23/2013	AM	Counseling
8	2S57ZI	Thompson, Alexandria B.	10/22/2013	AM	Vitals
9	2S57ZI	Thompson, Alexandria B.	10/22/2013	PM	Cholesterol
10	3VGRKR	Jones, Haley L.	10/23/2013	PM	Immunization

Defining Hash Objects That Process Multiple Sets of Data Items per Key Value

The two DECLARE statement argument tags MULTIDATA: and DUPLICATE: affect how SAS defines and fills a hash object that has multiple sets of data items per key value. Your DATA step produces different results depending on how you code these two argument tags.

Understanding the MULTIDATA Argument Tag

The MULTIDATA: argument tag can take one of two values: "YES" or "NO". The default value is "NO", and this setting does not allow multiple sets of data items per key value in the hash object. When you specify "YES", SAS allows multiple sets of data items per key value in the hash object.

When you add data that has multiple sets of data items per key value to a hash object and you do not specify MULTIDATA: "YES", SAS loads the set of data items associated with the first occurrence of a key value into the hash object and ignores all other sets of data items associated with that key value. SAS does not issue any notes or warnings that the data contained multiple sets of data items per key value and that it loaded only one set of data items into the hash object.

The examples in the previous chapters did not specify the MULTIDATA: argument tag and so none of them allowed multiple sets of data items per key value. Example 4.7 took advantage of this default action to find the unique set of IDs in a data set.

Understanding the DUPLICATE Argument Tag

The DUPLICATE: argument tag can take one of two values: "REPLACE" or "ERROR". Both of these options affect processing only when SAS loads a data set into a hash object and the MULTIDATA: "NO" option is in effect. When you specify the MULTIDATA: "YES" option, SAS ignores the DUPLICATE: option if you include it on the DECLARE statement.

The REPLACE option on the DUPLICATE: argument tag causes SAS to store the *last* set of data items for a key value. This is in contrast to the default action of MULTIDATA: "NO" where SAS stores the *first* set of data items for a key value. With DUPLICATE: REPLACE, each time SAS reads a new set of data items for a key, it overwrites an existing set of data items for that key value in the hash object.

The ERROR option on the DUPLICATE: argument tag causes SAS to set an error condition when it finds a second set of data items for a key value. SAS sets the automatic variable _ERROR_ to 1, writes an error message to the SAS log, and stops the DATA step. The ERROR option can be useful when you are not sure whether multiple sets of data items per key value exist, and that for your application to execute correctly, multiple sets of data items must not exist.

The syntax for these two optional argument tags on the DECLARE statement follows:

```
multidata: 'YES'|'Y'  'NO'|'N'

duplicate: 'REPLACE'|'R'  'ERROR'|'E'
```

Comparing the Features of the MULTIDATA and DUPLICATE Argument Tags

Table 5.1 describes how the combinations of the options of MULTIDATA: and DUPLICATE: work together.

Table 5.1 Features of the DECLARE Statement MULTIDATA and DUPLICATE Argument Tags

	DUPLICATE Argument Tag		
MULTIDATA Argument Tag	**Not Specified**	**"REPLACE"**	**"ERROR"**
"YES"	Multiple sets of data items per key value are allowed.	Multiple sets of data items per key value are allowed. SAS ignores the REPLACE option.	Multiple sets of data items per key value are allowed. SAS ignores the ERROR option.
"NO" / Not Specified	Multiple sets of data items per key value are not allowed. SAS loads into the hash object only the **first** set of data items that it reads for a key value.	Multiple sets of data items per key value are not allowed. SAS retains in the hash object only the **last** set of data items that it reads for a key value.	Multiple sets of data items per key value are not allowed. If an additional set of data items for a key value is found, SAS sets an error condition and stops the DATA step.

Illustrating How the MULTIDATA and DUPLICATE Argument Tags Affect Hash Object Processing

This section presents several DATA steps that specify different combinations of the options for the MULTIDATA and DUPLICATE argument tags on the DECLARE statement. Examples 5.1 through 5.4 illustrate how SAS handles the different options. Table 5.1 in the previous section summarizes the combinations.

All DATA steps in this section process data set BPMEASURES. This data set contains seven blood pressure measurements for one patient over a 4-day period. One measurement was recorded on May 13, three on May 14, two on May 15, and one on May 16. The time that the measurement was made is recorded in variable BPTIME. The observations are stored in chronological order. Variable PTID identifies the patient.

Output 5.1 lists the observations in BPMEASURES.

Output 5.1 PROC PRINT of BPMEASURES

Obs	ptid	bpdate	bptime	systol	diast
1	AU81750Y	05/13/2013	10:15 AM	131	79
2	AU81750Y	05/14/2013	8:30 AM	125	80
3	AU81750Y	05/14/2013	3:45 PM	141	83
4	AU81750Y	05/14/2013	7:10 PM	132	80
5	AU81750Y	05/15/2013	9:40 AM	125	73
6	AU81750Y	05/15/2013	1:45 PM	133	85
7	AU81750Y	05/16/2013	11:20 AM	128	78

Loading Only the First Set of Data Items per Key Value into a Hash Object

The goal of Example 5.1 is to load into a hash object only the first set of data items for a key value. This example loads data set BPMEASURES into hash object BP and specifies PTID and BPDATE as key items. The DECLARE statement does not specify either the MULTIDATA or the DUPLICATE argument tags. Therefore, if a patient has more than one measurement on a date, SAS loads into BP only the data from the first measurement that it reads for the patient on that date. Because the observations in data set BPMEASURES are ordered chronologically for each patient before Example 5.1 executes, the first set of data items for a key value corresponds to the first measurement for the day for the patient.

Since Example 5.1 includes the ORDERED: “YES” argument tag, the OUTPUT method retrieves entries from BP and writes observations to MULTDUP1 in order by the values of the two key variables, PTID and BPDATE.

Example 5.1 Loading Only the First Set of Data Items per Key Value into a Hash Object

```
data _null_;
  attrib ptid length=$8
         bpdate length=8 format=mmddyy10.
         bptime length=8 format=timeampm8.
         systol length=8 label='Systolic BP'
         diast  length=8 label='Diastolic BP';
  if _n_=1 then do;
    declare hash bp(dataset: 'bpmeasures', ordered: 'yes');
    bp.definekey('ptid','bpdate');
    bp.definedata('ptid','bpdate','bptime','systol','diast');
    bp.definedone();
    call missing(ptid,bpdate,bptime,systol,diast);
  end;

  rc=bp.output(dataset: 'multdup1');
run;
```

Output 5.2 shows the contents of MULTDUP1. Note that SAS entered only the first measurement for each day into hash object BP, and then output these entries to MULTDUP1.

Output 5.2 PROC PRINT of MULTDUP1

Obs	ptid	bpdate	bptime	systol	diast
1	AU81750Y	05/13/2013	10:15 AM	131	79
2	AU81750Y	05/14/2013	8:30 AM	125	80
3	AU81750Y	05/15/2013	9:40 AM	125	73
4	AU81750Y	05/16/2013	11:20 AM	128	78

Loading Multiple Sets of Data Items per Key Value into a Hash Object

The goal of Example 5.2 is to load all sets of data items for a key value into a hash object. Example 5.2 loads data set BPMEASURES into hash object BP, and specifies PTID and BPDATE as key items. Because a patient can have more than one measurement on one date, multiple sets of data items can exist in BPMEASURES.

With the MULTIDATA: “YES” argument tag specified, SAS allows multiple sets of data items per unique combination of PTID and BPDATE in hash object BP. SAS loads all of the observations from BPMEASURES into BP.

Since Example 5.2 includes the ORDERED: “YES” argument tag, the OUTPUT method retrieves entries from BP, and writes observations to MULTDUP2 in order by the values of the two key variables, PTID and BPDATE.

Example 5.2 Loading Multiple Sets of Data Items per Key Value

```
data _null_;
  attrib ptid length=$8
         bpdate length=8 format=mmddyy10.
         bptime length=8 format=timeampm8.
         systol length=8 label='Systolic BP'
         diast  length=8 label='Diastolic BP';
  if _n_=1 then do;
   declare hash bp(dataset: 'bpmeasures', ordered: 'yes', multidata: 'yes');
   bp.definekey('ptid','bpdate');
   bp.definedata('ptid','bpdate','bptime','systol','diast');
   bp.definedone();
   call missing(ptid,bpdate,bptime,systol,diast);
  end;

  rc=bp.output(dataset: 'multdup2');
run;
```

Output 5.3 shows the contents of MULTDUP2. SAS loads all seven observations from BPMEASURES into hash object BP, and outputs all entries to MULTDUP2.

Output 5.3 PROC PRINT of MULTDUP2

Obs	ptid	bpdate	bptime	systol	diast
1	AU81750Y	05/13/2013	10:15 AM	131	79
2	AU81750Y	05/14/2013	8:30 AM	125	80
3	AU81750Y	05/14/2013	3:45 PM	141	83
4	AU81750Y	05/14/2013	7:10 PM	132	80
5	AU81750Y	05/15/2013	9:40 AM	125	73
6	AU81750Y	05/15/2013	1:45 PM	133	85
7	AU81750Y	05/16/2013	11:20 AM	128	78

Retaining Only the Last Set of Data Items per Key Value in a Hash Object

The goal of Example 5.3 is to keep only the last set of data items in a hash object. This example loads data set BPMEASURES into hash object BP, and specifies PTID and BPDATE as key items. Because a patient can have more than one measurement on one date, multiple sets of data items can exist in BPMEASURES. If a patient has more than one measurement on a date, Example 5.3 keeps in BP only the data from the last measurement that it reads for the patient on each date.

The DECLARE statement in Example 5.3 does not specify the MULTIDATA argument tag, and it does specify the DUPLICATE: REPLACE argument tag. Without MULTIDATA: "YES" specified, SAS allows only one set of data items per key in the hash object.

The DUPLICATE: "REPLACE" argument tag causes SAS to overwrite each set of data items for a key value so that when SAS has finished processing all sets of data items for a key value, the *last* set of data items remains in the hash object. Because the observations in data set BPMEASURES are ordered chronologically before the DATA step executes, the last set of data items for a key value in Example 5.3 corresponds to the last measurement of the day.

Example 5.3 includes the ORDERED: "YES" argument tag so the OUTPUT method retrieves entries from BP, and writes observations to MULTDUP3 in order by the values of the two key variables, PTID and BPDATE.

Example 5.3 Retaining Only the Last Set of Data Items per Key Value in a Hash Object

```
data _null_;
  attrib ptid length=$8
         bpdate length=8 format=mmddyy10.
         bptime length=8 format=timeampm8.
         systol length=8 label='Systolic BP'
         diast  length=8 label='Diastolic BP';
  if _n_=1 then do;
    declare hash bp(dataset: 'bpmeasures', ordered: 'yes',
                    duplicate: 'replace');
    bp.definekey('ptid','bpdate');
    bp.definedata('ptid','bpdate','bptime','systol','diast');
    bp.definedone();
    call missing(ptid,bpdate,bptime,systol,diast);
  end;

  rc=bp.output(dataset: 'multdup3');
run;
```

Output 5.4 shows the contents of MULTDUP3. SAS overwrites data values so that only the last measurement for each day remains in hash object BP. SAS outputs these entries to MULTDUP3.

Output 5.4 PROC PRINT of MULTDUP3

Obs	ptid	bpdate	bptime	systol	diast
1	AU81750Y	05/13/2013	10:15 AM	131	79
2	AU81750Y	05/14/2013	7:10 PM	132	80
3	AU81750Y	05/15/2013	1:45 PM	133	85
4	AU81750Y	05/16/2013	11:20 AM	128	78

Preventing Creation of a Hash Object Where Multiple Sets of Data Items per Key Value Exist

The goal of Example 5.4 is to ensure that the data set being loaded into a hash object does not have multiple sets of data items per key value. This example loads data set BPMEASURES into hash object BP and specifies PTID and BPDATE as key items. Because a patient can have more than one measurement on one date, multiple sets of data items can exist in BPMEASURES.

The DECLARE statement specifies DUPLICATE: "ERROR". It does not specify the MULTIDATA: "YES" option, so SAS does not allow multiple sets of data items per key value in hash object BP. With the DUPLICATE: "ERROR" argument tag specified, the first time SAS encounters a key value with more than one set of data items, SAS sets an error condition and stops the DATA step. It does not identify any subsequent sets of data items for the key value currently being processed or for any of the remaining key values.

Example 5.4 Preventing Creation of a Hash Object Where Multiple Sets of Data Items per Key Value Exist

```
data _null_;
  attrib ptid length=$8
         bpdate length=8 format=mmddyy10.
         bptime length=8 format=timeampm8.
         systol length=8 label='Systolic BP'
         diast  length=8 label='Diastolic BP';
  if _n_=1 then do;
   declare hash bp(dataset: 'bpmeasures', ordered:'yes', duplicate:'error');
   bp.definekey('ptid','bpdate');
   bp.definedata('ptid','bpdate','bptime','systol','diast');
   bp.definedone();
   call missing(ptid,bpdate,bptime,systol,diast);
  end;

  rc=bp.output(dataset: 'multdup4');
run;
```

Since data set BPMEASURES contains multiple sets of data items for two combinations of the key values, the DATA step ends in error. SAS does not create data set MULTDUP4. The SAS log for Example 5.4 follows.

```
ERROR: Duplicate key found when loading data set bpmeasures at line
281 column 5.
NOTE: There were 3 observations read from the data set WORK.BPMEASURES.
ERROR: Hash data set load failed at line 281 column 5.
ERROR: DATA STEP Component Object failure.  Aborted during the
EXECUTION phase.
```

Understanding the Methods That Look for Multiple Sets of Data Items per Key

A typical retrieval of data from a hash object for a key value that has multiple sets of data items starts with executing the FIND method to find the key value and retrieve its first set of data items. After locating the key value, the FIND_NEXT or FIND_PREV methods look for additional sets of data items that have the same key value. The HAS_NEXT and HAS_PREV methods can determine if any additional sets of data items exist for the same key value, but these two methods do not return any data.

Finding Multiple Sets of Data Items

The FIND_NEXT and FIND_PREV methods find additional data items for the current key value. The FIND_NEXT method finds the next set of data items for the current key value while the FIND_PREV method finds the previous set of data items. When SAS finds another set of data items for the key value, the FIND_NEXT and FIND_PREV methods replace the values in the corresponding DATA step variables with the data from this new set of data items that it retrieved from the hash object. IF SAS does not find another set of data items for the key value, it does not alter any DATA step variable values.

The FIND_NEXT and FIND_PREV methods do not have any argument tags. These methods return a code that indicates whether the method found another set of data items for the current key value that was initially set by execution of the FIND method and whether the FIND_NEXT or FIND_PREV method returned the data item values to the corresponding data variables. A return code of 0 indicates that the method was successful, and a non-zero return code indicates that the method was not successful.

```
rc=object.FIND_NEXT();

rc=object.FIND_PREV();
```

Determining If Additional Sets of Data Items Exist

The HAS_NEXT and HAS_PREV methods check whether the current key value has any more sets of data items. The HAS_NEXT method looks forward for more data, and the HAS_PREV method looks backward. These two methods do not return any data items.

The RESULT: argument tag is the only argument tag that HAS_NEXT and HAS_PREV have, and SAS requires this argument tag on all applications of these two methods. The object of the RESULT: argument tag is the name of a DATA step variable. The action of RESULT: assigns a code to this variable that indicates whether another set of data items exists for the current key value. A value of 0 indicates that HAS_NEXT or HAS_PREV did not find any more sets of data items for the current key value. A non-zero value indicates that at least one more set of data items exists for the current key value in the direction that the specific method looks.

```
rc=object.HAS_NEXT(RESULT: variable-name);

rc=object.HAS_PREV(RESULT: variable-name);
```

The HAS_NEXT and HAS_PREV methods also set a return code to indicate whether the method executed. A value of 0 indicates that it did, while a non-zero value indicates that the method did not.

When writing code that checks execution of the HAS_NEXT and HAS_PREV methods, test the value of the variable that you assign to the RESULT: argument tag instead of the statement return code. The return code for these methods indicates only that the method executed, not necessarily that more data exists for the current key value.

The HAS_NEXT and HAS_PREV methods do not move through the sets of data items for a key value. Therefore, if you want to iterate through all sets of data items for a key value, you must follow a HAS_NEXT call with a FIND_NEXT call and follow a HAS_PREV call with a FIND_PREV call. For example, if you write your DO UNTIL or DO WHILE loop with the intention that it execute until it finds all sets of data items for a key value, and this loop had only a HAS_NEXT method call without a FIND_NEXT method call, you would have an infinite loop.

Understanding How SAS Stores Multiple Sets of Data Items per Key

When you add the ORDERED: "YES", ORDERED: "ASCENDING", or ORDERED: "DESCENDING" argument tag to your DECLARE statement, SAS retrieves data from a hash object in order by the values of the key items. The value of the ORDERED: argument tag affects only the order in which key values are retrieved from a hash object. It does not affect how SAS loads the multiple sets of data items for each key value into the hash object.

Starting in SAS 9.2 TS2M2, SAS retrieves multiple sets of data items from a hash object in the order they were encountered in the DATA step. This action is true even if you omit the ORDERED: "YES" argument tag. Knowing that SAS orders the multiple sets of items in this manner is useful when you apply the methods that examine each of the entries for a specific key value. Some of the examples in this chapter take advantage of this feature.

Looking back at Example 5.2, SAS loads all of the observations in data set BPMEASURES into hash object BP. BP is keyed by PTID and BPDATE. The observations in BPMEASURES are in order by PTID, BPDATE, and BPTIME when Example 5.2 executes. In SAS 9.2 TS2M2 and later, for each unique combination of PTID and BPDATE, SAS adds the data items to BP in chronological order since that is how they are arranged in data set BPMEASURES.

Because SAS does not alter the order of multiple sets of data items, it might be necessary for you to sort the data items before you load them into a hash object. While Example 5.2 does not require that you arrange the data items in chronological order, a different application might require that you sort the observations into the order you need them prior to the DATA step that defines the hash object.

Comparing Retrieval of Data from a Hash Object That Allows Multiple Sets of Data Items per Key Value to a Hash Iterator Object

Hash objects that allow multiple sets of data items per key value and hash iterator objects offer similar ways of processing data. The FIRST, LAST, NEXT, and PREV methods applied to a hash iterator object can traverse the underlying hash object in key order if the ORDERED: argument tag specifies that SAS retrieve data from the hash object in key order. The SETCUR method can specify a starting key value from which the traversal can progress.

Similar methods exist that allow movement within a key value for hash objects that have multiple sets of data items per key value. As described earlier in this chapter, methods FIND_NEXT and FIND_PREV move forward and backward through the multiple sets of data items for a key value, and methods HAS_NEXT and HAS_PREV evaluate whether more sets of data items exist for a key value.

Unless you define a hash iterator object and associate it with a hash object that has multiple sets of data items per key value, you must specify exact key values you want to examine in the hash object. You start by applying the FIND method to find the specific key value. Then you use the FIND_NEXT and FIND_PREV methods to examine the sets of data items for the specific key value.

If you know the key values that you want to find in a hash object that has multiple sets of data items per key value, it might be easier to write code that uses the methods that find specific key values instead of code that uses a hash iterator object to traverse the hash object. The methods that traverse a hash iterator object move either forward or backward through the hash object starting at the beginning, at the end, or at a specific key value. Retrieval from a hash object associated with a hash iterator object does not necessarily have to be in key value order. Either ordered retrieval or not, with a hash iterator solution, you would most likely have to add code that tests the key value returned from the hash object to determine when a key value changes, and you might need to add code to retain data between retrievals from the hash object.

Examples 5.5 and 5.6 accomplish the same task of finding data items for specific key values. Example 5.5 uses the methods specific to hash objects that can have multiple sets of data items per key value. In Example 5.6, a hash iterator object traverses an underlying hash object that does not allow multiple sets of data items per key value.

Examples 5.5 and 5.6 both use the data set described at the beginning of this chapter. Data set OCTOBEREVENT contains the schedule information for 52 employees assigned to work at a health screening event on October 22 and 23, 2013. Employees are assigned to a morning ("AM") or afternoon ("PM") shift to perform one of six tasks: cholesterol screening; counseling; dietary advice; immunization; pharmaceutical counseling; and measurement of vital health indicators such as blood pressure and weight.

The goal of Examples 5.5 and 5.6 is to find the scheduled shifts for four employees. Three of the four employees—OYMEE3, PTBHUP, and GWS2QS—have data in OCTOBEREVENT. One employee, K8821U, does not.

Output 5.5 shows the first 35 observations in OCTOBEREVENT. The three employees with data in OCTOBEREVENT have at least one row in Output 5.5.

Output 5.5 PROC PRINT of OCTOBEREVENT (first 35 observations)

Obs	eventdate	shift	activity	empid	empname
1	10/22/2013	AM	Cholesterol	02QYJG	Howard, Rachel Y.
2	10/22/2013	AM	Cholesterol	8I4YKY	Brooks, Shelby B.
3	10/22/2013	AM	Cholesterol	ADHW3A	Moore, Allison W.
4	10/22/2013	AM	Dietary	IFQZ8S	Nelson, Kelly U.
5	10/22/2013	AM	Dietary	OYMEE3	Thomas, Hannah I.
6	10/22/2013	AM	Dietary	PTBHUP	Thompson, Olivia P.
7	10/22/2013	AM	Pharmaceutical	WVV7PT	Patterson, Vanessa N.
8	10/22/2013	AM	Counseling	14ZN75	Miller, Sierra Q.
9	10/22/2013	AM	Counseling	5KA7JH	Martin, Aaron G.
10	10/22/2013	AM	Counseling	BA8CRZ	Baker, Marissa R.
11	10/22/2013	AM	Counseling	FAL5UZ	Butler, Haley P.
12	10/22/2013	AM	Counseling	TMJTVP	Edwards, Brittany O.
13	10/22/2013	AM	Counseling	YXW78P	Davis, Andrea Y.
14	10/22/2013	AM	Immunization	1CV01H	Brown, Lindsey V.
15	10/22/2013	AM	Immunization	8EHCSX	Hernandez, Elizabeth S.
16	10/22/2013	AM	Immunization	HD8ERT	Paine, Mike J.
17	10/22/2013	AM	Immunization	KRTV2T	Gonzalez, Alejandro F.
18	10/22/2013	AM	Vitals	NCPO4E	Richardson, Bailey U.
19	10/22/2013	AM	Pharmaceutical	P438U8	Price, Katelyn Q.
20	10/22/2013	AM	Vitals	2S57ZI	Thompson, Alexandria B.
21	10/22/2013	AM	Vitals	69BWGZ	Jones, Marissa W.
22	10/22/2013	AM	Vitals	F97SKU	Lee, Alicia Q.
23	10/22/2013	AM	Vitals	FTZMFI	Rodriguez, Samantha E.
24	10/22/2013	AM	Vitals	OV9K5X	Baker, Jasmine O.
25	10/22/2013	PM	Cholesterol	02QYJG	Howard, Rachel Y.
26	10/22/2013	PM	Cholesterol	2S57ZI	Thompson, Alexandria B.
27	10/22/2013	PM	Cholesterol	8I4YKY	Brooks, Shelby B.
28	10/22/2013	PM	Dietary	C800UH	Patterson, Jamie Q.
29	10/22/2013	PM	Dietary	GWS2QS	Parker, Brittany D.
30	10/22/2013	PM	Dietary	LISUUC	Davis, Laura W.
31	10/22/2013	PM	Counseling	OV9K5X	Baker, Jasmine O.
32	10/22/2013	PM	Counseling	OYMEE3	Thomas, Hannah I.
33	10/22/2013	PM	Counseling	WRH1W5	Chou, Alexandria U.
34	10/22/2013	PM	Counseling	4RD316	Cox, Christina Y.
35	10/22/2013	PM	Counseling	DVZ03Q	Ramirez, Alexa M.

Finding Sets of Data Items in a Hash Object That Allow Multiple Sets of Data Items per Key Value

The approach that Example 5.5 follows is to look for data for the four employees by searching a hash object that allows multiple sets of data items per key value. The DATA step starts by loading OCTOBEREVENT into hash object OE. The DECLARE statement includes the MULTIDATA: "YES" argument tag. The key for OE is EMPID.

A DO loop iterates through the list of employee IDs for whom schedule data is required. The first statement in the DO loop executes the FIND method to look in OE for the loop's current value of EMPID. If SAS finds the EMPID value, a DO UNTIL loop executes that outputs an observation for each set of data items present for that employee in OE. The last statement in the DO UNTIL loop calls the FIND_NEXT method, which looks for the next set of data items for the current value of EMPID.

When SAS does not find an EMPID value in OE, a DO block executes that assigns default text to variable ACTIVITY, and it also sets the other variables to missing.

Example 5.5 Finding Sets of Data Items in a Hash Object That Allow Multiple Sets of Data Items per Key Value

```
data sched4emps;
  attrib empid length=$6 label='Employee ID'
         empname length=$40 label='Employee Name'
         eventdate format=mmddyy10.
         shift length=$2 label='Event Shift'
         activity length=$20 label='Event Activity';

  declare hash oe(dataset: 'octoberevent', multidata: 'yes', ordered: 'yes');
  oe.definekey('empid');
  oe.definedata('eventdate','shift','activity','empname');
  oe.definedone();
  call missing(eventdate,shift,activity,empname);

  drop rc;
  do empid='OYMEE3','PTBHUP','GWS2QS','K8821U';
    rc=oe.find();
    if rc=0 then do;
      do until(rc ne 0);
        output;
        rc=oe.find_next();
      end;
    end;
    else do;
      activity='** Not Scheduled';
      call missing(eventdate,shift,empname);
      output;
    end;
  end;
run;
```

Traversing a Hash Object Using a Hash Iterator Object to Find Sets of Data Items per Key Value

Example 5.6 uses a hash iterator object to accomplish the same task that Example 5.5 accomplishes. Example 5.6 creates hash object OE2 from data set OCTOBEREVENT. Hash object OE2 does not allow multiple sets of data items per key value. Instead, this hash object defines three key items that uniquely define an observation in OCTOBEREVENT. Hash iterator object OEITER defines OE2 as its underlying hash object.

The DATA step defines two temporary arrays. Array FINDEMPID contains the list of EMPID values to search for in OE2. Array FOUND contains the same number of elements as FINDEMPID, and these elements track whether SAS found the associated EMPID value in OE2.

Traversal of OE2 starts at the first entry in OE2 by calling the FIRST method. A DO UNTIL loop looks at each returned key value from OE2. Variable IDX specifies the element number in FINDEMPID for the current value of EMPID. If the current value of EMPID does not equal any of the values in the FINDEMPID array, the WHICHC function assigns 0 to IDX. When SAS finds the value of EMPID in the FINDEMPID array, it sets the corresponding element in the FOUND array to 1 and outputs an observation.

When the DO UNTIL loop ends and SAS has examined all key values in OE2, an iterative DO loop executes that examines the values in the FOUND array. An element in FOUND that does not equal 1 means that the associated EMPID value is not present in OE2 and that the employee is not scheduled to work at the event. SAS assigns default text to ACTIVITY and sets variables EVENTDATE, SHIFT, and EMPNAME to missing.

Example 5.6 Using a Hash Iterator Object to Find Sets of Data Items per Key Value

```
data sched4emps;
  attrib empid length=$6 label='Employee ID'
         empname length=$40 label='Employee Name'
         eventdate format=mmddyy10.
         shift length=$2 label='Event Shift'
         activity length=$20 label='Event Activity';

  declare hash oe2(dataset: 'octoberevent', ordered: 'yes');
  declare hiter oeiter('oe2');
  oe2.definekey('empid','eventdate','shift');
  oe2.definedata('empid','eventdate','shift','activity','empname');
  oe2.definedone();
  call missing(empid,eventdate,shift,activity,empname);

  drop rc idx i;

  array findempid{4} $ 6 _temporary_
                      ('OYMEE3','PTBHUP','GWS2QS','K8821U');
  array found{4} _temporary_ ;
```

```
    rc=oeiter.first();
    do until(rc ne 0);
      idx=whichc(empid,of findempid[*]);
      if idx gt 0 then do;
        found{idx}=1;
        output;
      end;
      rc=oeiter.next();
    end;
    do i=1 to dim(findempid);
      if found{i} ne 1 then do;
        empid=findempid{i};
        activity='** Not Scheduled';
        call missing(eventdate,shift,empname);
        output;
      end;
    end;
  run;
```

Comparing the Data Sets Created by Example 5.5 and by Example 5.6

Examples 5.5 and 5.6 create the same data set. However, the observations in the two data sets that the two examples create are not in the same order. The observations in SCHED4EMPS created by Example 5.5 are in order by the arrangement of the EMPID values on the DO EMPID= statement. The observations in SCHED4EMPS created by Example 5.6 are in order by the key items on the DEFINEKEY statement for the EMPID values that SAS finds in hash object OE2 with an observation for each of the EMPID values that it does not find appended at the end.

Output 5.6 shows the contents of data set SCHED4EMPS with the observations in the order that Example 5.5 created.

Output 5.6 PROC PRINT of SCHED4EMPS As Produced by Example 5.5

Obs	empid	empname	eventdate	shift	activity
1	OYMEE3	Thomas, Hannah I.	10/22/2013	AM	Dietary
2	OYMEE3	Thomas, Hannah I.	10/22/2013	PM	Counseling
3	OYMEE3	Thomas, Hannah I.	10/23/2013	PM	Dietary
4	PTBHUP	Thompson, Olivia P.	10/22/2013	AM	Dietary
5	PTBHUP	Thompson, Olivia P.	10/22/2013	PM	Immunization
6	GWS2QS	Parker, Brittany D.	10/22/2013	PM	Dietary
7	GWS2QS	Parker, Brittany D.	10/23/2013	AM	Cholesterol
8	GWS2QS	Parker, Brittany D.	10/23/2013	PM	Vitals
9	K8821U		.		** Not Scheduled

Modifying Data in a Hash Object That Allows Multiple Sets of Data Items per Key Value

Chapter 4 described the ADD and REMOVE methods that can modify the contents of a hash object that allows multiple sets of data items per key. Chapter 4 also described the REPLACE method. However, SAS did not design the REPLACE method to work with hash objects that allow multiple sets of data items per key value. This section will not present any examples that use REPLACE.

The ADD method is specified in the same way as shown in Chapter 4 where multiple sets of data items were not allowed. As long as you define the hash object with the MULTIDATA: "YES" argument, you can add additional sets of data items to the hash object for a specific key value.

The REMOVE method is also specified in the same way as was shown in Chapter 4. However, when the MULTIDATA: "YES" argument tag is specified, this method removes *all* sets of data items associated with the key value, not just a single set of data items.

When you want to remove or replace *only* the current set of data items, you can use the REMOVEDUP and REPLACEDUP methods. First, you must use the FIND, FIND_NEXT, or FIND_PREV methods to locate the specific set of data items you want to remove or replace. Second, you apply the REMOVEDUP or REPLACEDUP method.

The optional KEY: argument tags in the REMOVEDUP method are specified the same way as they are with the REMOVE method. However, the REMOVEDUP method removes only the current set of data items for the specific key value.

The REPLACEDUP method has only DATA: argument tags, and these tags are optional. Since REPLACEDUP does not have KEY: argument tags, your code must first find the correct key value, and then next find the correct set of data items for the key value.

```
rc=object.REMOVEDUP(<KEY: keyvalue-1,..., KEY: keyvalue-n>);

rc=object.REPLACEDUP(<DATA: datavalue-1,..., DATA: datavalue-n>);
```

When the REMOVEDUP method executes, you might expect SAS to move its pointers to the next set or previous set of items that has the same key value. However, SAS instead sets the pointers to null. Therefore, you would need to start over searching for the same key value by issuing a FIND method call, and possibly to follow that with FIND_NEXT or FIND_PREV method calls.

Examples 5.7 through 5.10 illustrate how these methods modify data in a hash object that allows multiple sets of data items per key value. All of the data sets use data set OCTOBEREVENT. This data set contains the schedule information for 52 employees scheduled to work at a health screening event on October 22 and 23, 2013. Employees are assigned to a morning ("AM") or afternoon ("PM") shift to perform one of six tasks: cholesterol screening; counseling; dietary advice; immunization; pharmaceutical counseling; and measurement of vital health indicators such as blood pressure and weight. Several employees work more than one shift. Data set

OCTOBEREVENT contains 105 observations. Output 5.5 lists the first 35 observations in OCTOBEREVENT.

Adding Data to a Hash Object That Allows Multiple Sets of Data Items per Key Value

Example 5.7 reads in two new observations and adds data from these observations to a hash object. Example 5.7 loads data set OCTOBEREVENT into hash object MULT1, and specifies that MULT1 can have multiple sets of data items per key value. The two new observations contain schedule information for two employees on October 23 in the morning.

The key items in MULT1 are EVENTDATE, SHIFT, and ACTIVITY. Because more than one employee can work a shift and activity on a specific date, multiple values of EMPID and EMPNAME can exist for each unique combination of EVENTDATE, SHIFT, and ACTIVITY.

Output 5.7 shows the observations in OCTOBEREVENT for the October 23 morning schedule.

Output 5.7 PROC PRINT of OCTOBEREVENT for Observations Where EVENTDATE='23OCT2013'D and SHIFT='AM'

Obs	eventdate	shift	activity	empid	empname
50	10/23/2013	AM	Cholesterol	4RD316	Cox, Christina Y.
51	10/23/2013	AM	Cholesterol	GWS2QS	Parker, Brittany D.
52	10/23/2013	AM	Cholesterol	HD8ERT	Paine, Mike J.
53	10/23/2013	AM	Dietary	IEZF48	Thompson, Brianna I.
54	10/23/2013	AM	Dietary	K42ODX	Anderson, Rebecca J.
55	10/23/2013	AM	Pharmaceutical	NGRZ6L	Coleman, Madeline I.
56	10/23/2013	AM	Pharmaceutical	OV9K5X	Baker, Jasmine O.
57	10/23/2013	AM	Counseling	1CV01H	Brown, Lindsey V.
58	10/23/2013	AM	Counseling	81DJ0Z	Martinez, Kelly I.
59	10/23/2013	AM	Counseling	ADHW3A	Moore, Allison W.
60	10/23/2013	AM	Counseling	FTZMFI	Rodriguez, Samantha E.
61	10/23/2013	AM	Counseling	INZOIM	Mitchell, Olivia M.
62	10/23/2013	AM	Counseling	TMJTVP	Edwards, Brittany O.
63	10/23/2013	AM	Counseling	WVV7PT	Patterson, Vanessa N.
64	10/23/2013	AM	Counseling	02QYJG	Howard, Rachel Y.
65	10/23/2013	AM	Immunization	67Z8K3	Moore, Vanessa I.
66	10/23/2013	AM	Immunization	8I4YKY	Brooks, Shelby B.
67	10/23/2013	AM	Immunization	ICDUYL	Hill, Samantha S.
68	10/23/2013	AM	Immunization	P438U8	Price, Katelyn Q.
69	10/23/2013	AM	Immunization	VUH0Z4	Jones, Catherine H.

Obs	eventdate	shift	activity	empid	empname
70	10/23/2013	AM	Vitals	WRH1W5	Chou, Alexandria U.
71	10/23/2013	AM	Immunization	YU01P0	Mitchell, Catherine A.
72	10/23/2013	AM	Vitals	70M6YN	King, Mary K.
73	10/23/2013	AM	Vitals	8VLUT7	Hall, Kelsey H.
74	10/23/2013	AM	Vitals	BA8CRZ	Baker, Marissa R.
75	10/23/2013	AM	Vitals	FAL5UZ	Butler, Haley P.
76	10/23/2013	AM	Vitals	IY69JA	Martinez, Lauren B.
77	10/23/2013	AM	Vitals	V1DPL5	Diaz, Lindsey N.

Example 5.7 reads the raw data, and then adds data from these new observations to hash object MULT1. The last action in the DATA step outputs the contents of MULT1 to data set OCTOBEREVENT_REV1. This new data set contains the 105 observations in OCTOBEREVENT plus the two new observations that the DATA step read and added to MULT1.

The two new observations are shown following the DATALINES statement in Example 5.7.

Example 5.7 Adding Data to a Hash Object That Allows Multiple Sets of Data Items per Key Value

```
data _null_;
  attrib eventdate format=mmddyy10.
         shift length=$2 label='Event Shift'
         activity length=$20 label='Event Activity'
         empid length=$6 label='Employee ID'
         empname length=$40 label='Employee Name';

  if _n_=1 then do;
    declare hash mult1(dataset: 'octoberevent',multidata:
                       'yes',ordered: 'yes');
    mult1.definekey('eventdate','shift','activity');
    mult1.definedata('eventdate','shift','activity','empid',
                     'empname');
    mult1.definedone();

    call missing(eventdate,shift,activity,empid,empname);
  end;
  infile datalines eof=done;
  input @1 eventdate mmddyy8. @10 shift $2.
        @13 activity $14. @29 empid $6. @37 empname $40.;
  rc=mult1.add();
  return;
  done:
    rc=mult1.output(dataset: 'octoberevent_rev1');
```

```
datalines;
10232013 AM Cholesterol      Z1EJD9  Howard, Amanda X.
10232013 AM Dietary          YH2W10  Li, Louise L.
run;
```

Output 5.8 shows the observations in OCTOBEREVENT_REV1 for the October 23 morning shift. The two new observations are in **bold** in OUTPUT 5.8. Because hash object MULT1 was defined as ORDERED: 'YES', the OUTPUT method writes observations to data set OCTOBEREVENT_REV1 in order by the key items' values.

Employee Z1EJD9 has other data in input data set OCTOBEREVENT. Employee YH2W10 is new.

Output 5.8 PROC PRINT of OCTOBEREVENT_REV1 for Observations where EVENTDATE='23OCT2013'D and SHIFT='AM'

Obs	eventdate	shift	activity	empid	empname
50	10/23/2013	AM	Cholesterol	4RD316	Cox, Christina Y.
51	10/23/2013	AM	Cholesterol	GWS2QS	Parker, Brittany D.
52	10/23/2013	AM	Cholesterol	HD8ERT	Paine, Mike J.
53	**10/23/2013**	**AM**	**Cholesterol**	**Z1EJD9**	**Howard, Amanda X.**
54	10/23/2013	AM	Counseling	1CV01H	Brown, Lindsey V.
55	10/23/2013	AM	Counseling	81DJ0Z	Martinez, Kelly I.
56	10/23/2013	AM	Counseling	ADHW3A	Moore, Allison W.
57	10/23/2013	AM	Counseling	FTZMFI	Rodriguez, Samantha E.
58	10/23/2013	AM	Counseling	INZOIM	Mitchell, Olivia M.
59	10/23/2013	AM	Counseling	TMJTVP	Edwards, Brittany O.
60	10/23/2013	AM	Counseling	WVV7PT	Patterson, Vanessa N.
61	10/23/2013	AM	Counseling	02QYJG	Howard, Rachel Y.
62	10/23/2013	AM	Dietary	IEZF48	Thompson, Brianna I.
63	10/23/2013	AM	Dietary	K42ODX	Anderson, Rebecca J.
64	**10/23/2013**	**AM**	**Dietary**	**YH2W10**	**Li, Louise L.**
65	10/23/2013	AM	Immunization	67Z8K3	Moore, Vanessa I.
66	10/23/2013	AM	Immunization	8I4YKY	Brooks, Shelby B.
67	10/23/2013	AM	Immunization	ICDUYL	Hill, Samantha S.
68	10/23/2013	AM	Immunization	P438U8	Price, Katelyn Q.
69	10/23/2013	AM	Immunization	VUH0Z4	Jones, Catherine H.
70	10/23/2013	AM	Immunization	YU01P0	Mitchell, Catherine A.
71	10/23/2013	AM	Pharmaceutical	NGRZ6L	Coleman, Madeline I.
72	10/23/2013	AM	Pharmaceutical	OV9K5X	Baker, Jasmine O.
73	10/23/2013	AM	Vitals	WRH1W5	Chou, Alexandria U.

Obs	eventdate	shift	activity	empid	empname
74	10/23/2013	AM	Vitals	70M6YN	King, Mary K.
75	10/23/2013	AM	Vitals	8VLUT7	Hall, Kelsey H.
76	10/23/2013	AM	Vitals	BA8CRZ	Baker, Marissa R.
77	10/23/2013	AM	Vitals	FAL5UZ	Butler, Haley P.
78	10/23/2013	AM	Vitals	IY69JA	Martinez, Lauren B.
79	10/23/2013	AM	Vitals	V1DPL5	Diaz, Lindsey N.

Removing All Sets of Data Items for a Specific Key

Example 5.8 removes all data items for a specific key in a hash object that allows multiple sets of data items per key value. The DATA step loads data set OCTOBEREVENT into hash object MULT2 and specifies that MULT2 can have multiple sets of data items per key value. The goal of Example 5.8 is to remove all afternoon dietary activity events.

Output 5.9 includes the six observations for the afternoon dietary activity on both days.

Output 5.9 PROC PRINT of OCTOBEREVENT That Includes Observations Where SHIFT='PM' and ACTIVITY='DIETARY'

Obs	eventdate	shift	activity	empid	empname
	...				
26	10/22/2013	PM	Cholesterol	2S57ZI	Thompson, Alexandria B.
27	10/22/2013	PM	Cholesterol	8I4YKY	Brooks, Shelby B.
28	10/22/2013	PM	Dietary	C800UH	Patterson, Jamie Q.
29	10/22/2013	PM	Dietary	GWS2QS	Parker, Brittany D.
30	10/22/2013	PM	Dietary	LISUUC	Davis, Laura W.
31	10/22/2013	PM	Counseling	OV9K5X	Baker, Jasmine O.
32	10/22/2013	PM	Counseling	OYMEE3	Thomas, Hannah I.
	...				
79	10/23/2013	PM	Cholesterol	ADHW3A	Moore, Allison W.
80	10/23/2013	PM	Cholesterol	F97SKU	Lee, Alicia Q.
81	10/23/2013	PM	Dietary	IEZF48	Thompson, Brianna I.
82	10/23/2013	PM	Dietary	JDFPHH	Roberts, Laura U.
83	10/23/2013	PM	Dietary	OYMEE3	Thomas, Hannah I.
84	10/23/2013	PM	Counseling	WRH1W5	Chou, Alexandria U.
85	10/23/2013	PM	Pharmaceutical	YXW78P	Davis, Andrea Y.
	...				

A simple DATA step that uses a WHERE statement instead of a hash object can easily achieve the same outcome as Example 5.8. However, the point of Example 5.8 is to show how the REMOVE method functions when working with multiple sets of data items per key value. You would most likely find the most use in adapting Example 5.8 for your applications when the changes you need to make to your hash object are more complex than just removing entries.

Hash object MULT2 is keyed by SHIFT and ACTIVITY, the two variables whose values determine the entries to remove from MULT2. Since more than one employee can work at a dietary shift and since the dietary activity is held on both days, multiple values of EMPID, EMPNAME, and EVENTDATE can exist for each unique combination of SHIFT and ACTIVITY. A single call to the REMOVE method removes *all* of the data items associated with a unique combination of SHIFT and ACTIVITY.

The last action in the DATA step outputs the contents of MULT2 to data set OCTOBEREVENT_REV2. This new data set contains 99 observations. The six observations that correspond to the dietary activity in the afternoons of both days are not in OCTOBEREVENT_REV2.

Example 5.8 Removing All Sets of Data Items for a Specific Key

```
data _null_;
  attrib eventdate format=mmddyy10.
         shift length=$2 label='Event Shift'
         activity length=$20 label='Event Activity'
         empid length=$6 label='Employee ID'
         empname length=$40 label='Employee Name';

  declare hash mult2(dataset: 'octoberevent',multidata: 'yes',
                     ordered: 'yes');
  mult2.definekey('shift','activity');
  mult2.definedata('eventdate','shift','activity','empid','empname');
  mult2.definedone();

  call missing(eventdate,shift,activity,empid,empname);

  rc=mult2.remove(key:'PM',key:'Dietary');
  rc=mult2.output(dataset: 'octoberevent_rev2');
run;
```

Since MULTI2 is ordered by SHIFT and ACTIVITY, output data set OCTOBEREVENT_REV2 is ordered by SHIFT and ACTIVITY. Output 5.10 shows that the afternoon dietary activities have been removed.

Output 5.10 PROC PRINT of OCTOBEREVENT_REV2 for Selected Observations Where SHIFT='PM'

Obs	eventdate	shift	activity	empid	empname
63	10/22/2013	PM	Counseling	DVZ03Q	Ramirez, Alexa M.
64	10/22/2013	PM	Counseling	NGRZ6L	Coleman, Madeline I.
65	10/23/2013	PM	Counseling	WRH1W5	Chou, Alexandria U.
66	10/23/2013	PM	Counseling	67Z8K3	Moore, Vanessa I.
67	10/23/2013	PM	Counseling	K42ODX	Anderson, Rebecca J.
68	10/23/2013	PM	Counseling	KRTV2T	Gonzalez, Alejandro F.
69	10/23/2013	PM	Counseling	TMJTVP	Edwards, Brittany O.
70	10/23/2013	PM	Counseling	V1DPL5	Diaz, Lindsey N.
71	10/23/2013	PM	Counseling	Z1EJD9	Howard, Amanda X.
72	10/22/2013	PM	Immunization	67Z8K3	Moore, Vanessa I.
73	10/22/2013	PM	Immunization	8VLUT7	Hall, Kelsey H.
74	10/22/2013	PM	Immunization	PTBHUP	Thompson, Olivia P.
75	10/23/2013	PM	Immunization	3VGRKR	Jones, Haley L.
76	10/23/2013	PM	Immunization	69BWGZ	Jones, Marissa W.
77	10/23/2013	PM	Immunization	81DJ0Z	Martinez, Kelly I.
78	10/23/2013	PM	Immunization	NGRZ6L	Coleman, Madeline I.
79	10/23/2013	PM	Immunization	YU01P0	Mitchell, Catherine A.
80	10/22/2013	PM	Pharmaceutical	R89AM0	Baker, Amy R.

Removing Specific Sets of Data Items

Example 5.9 removes specific sets of data items for a specific key value in a hash object that allows multiple sets of data items per key value. The DATA step loads data set OCTOBEREVENT into hash object MULT3 and specifies that MULT3 can have multiple sets of data items per key value.

The goal of Example 5.9 is to modify the schedule on October 22 for the Counseling activity. Only three employees are needed in the morning shift and only two in the afternoon, and in OCTOBEREVENT six employees are scheduled for each shift. Hash object MULT3 is keyed by EVENTDATE, SHIFT, and ACTIVITY, which are the three variables whose values determine the entries to remove from MULT3.

Output 5.11 lists the observations in OCTOBEREVENT for the Counseling activity on October 22.

Output 5.11 PROC PRINT of OCTOBEREVENT That Includes Observations for the Counseling Activity on October 22

Obs	eventdate	shift	activity	empid	empname
8	10/22/2013	AM	Counseling	14ZN75	Miller, Sierra Q.
9	10/22/2013	AM	Counseling	5KA7JH	Martin, Aaron G.
10	10/22/2013	AM	Counseling	BA8CRZ	Baker, Marissa R.
11	10/22/2013	AM	Counseling	FAL5UZ	Butler, Haley P.
12	10/22/2013	AM	Counseling	TMJTVP	Edwards, Brittany O.
13	10/22/2013	AM	Counseling	YXW78P	Davis, Andrea Y.
31	10/22/2013	PM	Counseling	OV9K5X	Baker, Jasmine O.
32	10/22/2013	PM	Counseling	OYMEE3	Thomas, Hannah I.
33	10/22/2013	PM	Counseling	WRH1W5	Chou, Alexandria U.
34	10/22/2013	PM	Counseling	4RD316	Cox, Christina Y.
35	10/22/2013	PM	Counseling	DVZ03Q	Ramirez, Alexa M.
36	10/22/2013	PM	Counseling	NGRZ6L	Coleman, Madeline I.

After defining hash object MULT3 and loading it with data, the iterative DO loop in Example 5.9 examines the entries in MULT3. The iterative DO loop executes twice, once for the morning Counseling shift on October 22, and once for the afternoon Counseling shift on October 22. Assignment statements initialize variables NEMPTS and SHIFTMAX at the top of the loop. Variable NEMPS tallies the number of employees scheduled in each shift. Variable SHIFTMAX identifies the total number of employees needed in the shift.

The FIND method finds the first Counseling employee on October 22 for the current value of SHIFT. The DO UNTIL loop executes the FIND_NEXT method to find all employees scheduled to work the shift and the accumulator statement tallies the number of employees scheduled to work the shift.

If the value of NEMPS is greater than the number needed to work the shift, the inner iterative DO loop executes the number of times equal to the number of employees to drop from the schedule. Since six employees are scheduled to work each of the shifts, the program drops from the schedule the first three employees in the morning shift and the first four employees in the afternoon shift.

The FIND method executes on each iteration of the DO SHIFT= loop, and this action sets the pointers to the first employee with the key values. The REMOVEDUP method removes this first employee's entry from the hash object. Execution of REMOVEDUP sets to null the pointers that would position SAS at the next entry so you must start over for each entry you want to remove from MULT3 by invoking the FIND method.

The last action in the DATA step outputs the contents of MULT3 to data set OCTOBEREVENT_REV3. This new data set contains 98 observations.

Example 5.9 Removing Specific Sets of Data Items

```
data _null_;
  attrib eventdate format=mmddyy10.
         shift length=$2 label='Event Shift'
         activity length=$20 label='Event Activity'
         empid length=$6 label='Employee ID'
         empname length=$40 label='Employee Name';

  declare hash mult3(dataset: 'octoberevent',multidata: 'yes',
                     ordered: 'yes');
  mult3.definekey('eventdate','shift','activity');
  mult3.definedata('empid','empname','eventdate','shift','activity');
  mult3.definedone();

  call missing(eventdate,shift,activity,empid,empname);

  do shift='AM','PM';
    nemps=0;
    if shift='AM' then shiftmax=3;
    else if shift='PM' then shiftmax=2;
    rc=mult3.find(key: '22oct2013'd, key: shift, key: 'Counseling');
    do until (rc ne 0);
      nemps+1;
      rc=mult3.find_next();
    end;
    if nemps gt shiftmax then do;
      do i=1 to nemps-shiftmax;
        rc=mult3.find(key: '22oct2013'd, key: shift,
                      key: 'Counseling');
        rc=mult3.removedup();
      end;
    end;
  end;
  rc=mult3.output(dataset: 'octoberevent_rev3');
run;
```

Data set OCTOBEREVENT_REV3 has three employees scheduled to work the Counseling activity on October 22 in the morning, and two employees to work in the afternoon. Output 5.12 lists these employees' observations.

Output 5.12 PROC PRINT of OCTOBEREVENT_REV3 That Includes Observations for the Counseling Activity on October 22

Obs	empid	empname	eventdate	shift	activity
1	02QYJG	Howard, Rachel Y.	10/22/2013	AM	Cholesterol
2	8I4YKY	Brooks, Shelby B.	10/22/2013	AM	Cholesterol
3	ADHW3A	Moore, Allison W.	10/22/2013	AM	Cholesterol
4	FAL5UZ	Butler, Haley P.	10/22/2013	AM	Counseling
5	TMJTVP	Edwards, Brittany O.	10/22/2013	AM	Counseling
6	YXW78P	Davis, Andrea Y.	10/22/2013	AM	Counseling
7	IFQZ8S	Nelson, Kelly U.	10/22/2013	AM	Dietary
	. . .				
20	FTZMFI	Rodriguez, Samantha E.	10/22/2013	AM	Vitals
21	OV9K5X	Baker, Jasmine O.	10/22/2013	AM	Vitals
22	02QYJG	Howard, Rachel Y.	10/22/2013	PM	Cholesterol
23	2S57ZI	Thompson, Alexandria B.	10/22/2013	PM	Cholesterol
24	8I4YKY	Brooks, Shelby B.	10/22/2013	PM	Cholesterol
25	DVZ03Q	Ramirez, Alexa M.	10/22/2013	PM	Counseling
26	NGRZ6L	Coleman, Madeline I.	10/22/2013	PM	Counseling
27	C800UH	Patterson, Jamie Q.	10/22/2013	PM	Dietary
28	GWS2QS	Parker, Brittany D.	10/22/2013	PM	Dietary
	. . .				

Replacing Data in Specific Sets of Data Items

Example 5.10 replaces data in specific sets of data items for specific key values in a hash object that allows multiple sets of data items per key value. The DATA step loads data set OCTOBEREVENT into hash object MULT4 and specifies that MULT4 can have multiple sets of data items per key value. The key in MULT4 is EMPID since the modifications Example 5.10 makes are based on values of EMPID.

Example 5.10 has three goals:

- For EMPID= "02QYJG", change the morning activities to "Immunization" and the afternoon activities to "Vitals". This employee is scheduled for "Cholesterol" on the October 22 morning and afternoon shifts and "Counseling" on the October 23 afternoon shift.
- Reassign afternoon activities for EMPID= "ADHW3A" to new employee 724RTQ. Employee ADHW3A is scheduled for all shifts on October 22 and October 23.
- Output the revised data set in order by the event date, shift, activity, and employee ID.

The PROC PRINT report in Output 5.13 includes all of the observations for employees 02QYJG and ADHW3A in input data set OCTOBEREVENT. The observations for these two employees are in **bold**.

Output 5.13 PROC PRINT of OCTOBEREVENT (includes all observations for EMPID="02QYJG" and EMPID="ADHW3A"

Obs	eventdate	shift	activity	empid	empname
1	**10/22/2013**	**AM**	**Cholesterol**	**02QYJG**	**Howard, Rachel Y.**
2	10/22/2013	AM	Cholesterol	8I4YKY	Brooks, Shelby B.
3	**10/22/2013**	**AM**	**Cholesterol**	**ADHW3A**	**Moore, Allison W.**
4	10/22/2013	AM	Dietary	IFQZ8S	Nelson, Kelly U.
5	10/22/2013	AM	Dietary	OYMEE3	Thomas, Hannah I.
	. . .				
23	10/22/2013	AM	Vitals	FTZMFI	Rodriguez, Samantha E.
24	10/22/2013	AM	Vitals	OV9K5X	Baker, Jasmine O.
25	**10/22/2013**	**PM**	**Cholesterol**	**02QYJG**	**Howard, Rachel Y.**
26	10/22/2013	PM	Cholesterol	2S57ZI	Thompson, Alexandria B.
	. . .				
43	10/22/2013	PM	Vitals	7YN5NV	Perez, Paige O.
44	**10/22/2013**	**PM**	**Vitals**	**ADHW3A**	**Moore, Allison W.**
45	10/22/2013	PM	Vitals	HD8ERT	Paine, Mike J.
	. . .				
58	10/23/2013	AM	Counseling	81DJ0Z	Martinez, Kelly I.
59	**10/23/2013**	**AM**	**Counseling**	**ADHW3A**	**Moore, Allison W.**
60	10/23/2013	AM	Counseling	FTZMFI	Rodriguez, Samantha E.
61	10/23/2013	AM	Counseling	INZOIM	Mitchell, Olivia M.
62	10/23/2013	AM	Counseling	TMJTVP	Edwards, Brittany O.
63	10/23/2013	AM	Counseling	WVV7PT	Patterson, Vanessa N.
64	**10/23/2013**	**AM**	**Counseling**	**02QYJG**	**Howard, Rachel Y.**
65	10/23/2013	AM	Immunization	67Z8K3	Moore, Vanessa I.
	. . .				
78	10/23/2013	PM	Cholesterol	7YN5NV	Perez, Paige O.
79	**10/23/2013**	**PM**	**Cholesterol**	**ADHW3A**	**Moore, Allison W.**
80	10/23/2013	PM	Cholesterol	F97SKU	Lee, Alicia Q.
	. . .				

Example 5.10 requires two hash objects to complete its three goals. SAS modifies data items in the first hash object and outputs the contents of the first hash object to a data set. SAS loads the

data set that the first hash object created into a second hash object. SAS outputs the second hash object to a newer version of the data set in order by the values of the key items.

Processing the First Hash Object in Example 5.10

SAS loads data set OCTOBEREVENT into the first hash object MULT4. This hash object allows multiple sets of data items per key value, and SAS retrieves the entries from MULT4 in order by the values of its key item EMPID. Instances of the REPLACEDUP method modify the contents of MULT4.

SAS language statements modify MULT4. Once all changes have been made, the OUTPUT method outputs the contents of MULT4 to data set OCTOBEREVENT_REV4. SAS orders the observations in this data set by the values of EMPID. The observations are not yet in order by the event date, shift, activity, and employee ID that the application requires. SAS retrieves data from the second hash object OERP, which is described below, in the order required.

The first FIND method call finds the first occurrence in MULT4 for employee 02QYJG. The first DO UNTIL loop then examines each set of data items for this employee and changes this employee's morning and afternoon activities. The call to REPLACEDUP does not include any argument tags. All of the data items except for ACTIVITY remain the same. The IF-THEN statements change the value of DATA step variable ACTIVITY so that when REPLACEDUP executes, this new value replaces the original value for the ACTIVITY data item in MULT4.

The second call to the FIND method finds the first occurrence in MULT4 for employee ADHW3A. The second DO UNTIL loop examines each set of data items for this second employee. When SHIFT="PM", SAS assigns new values to data items EMPID and EMPNAME. The call to REPLACEDUP does not include any argument tags. All of the data items except for EMPID and EMPNAME remain the same. When REPLACEDUP executes, the new values for EMPID and EMPNAME replace the original values in MULT4.

Processing the Second Hash Object in Example 5.10

PROC SORT could be called to sort the observations in OCTOBEREVENT_REV4 after the DATA step finishes. Instead, Example 5.10 loads a second hash object, OERP, with the observations from OCTOBEREVENT_REV4 and specifies that SAS retrieve data from OERP in order by the values of the four key items that define the required sort order of the final data set. The combinations of the values of these key items form unique entries in OERP, so the definition of OERP omits the MULTIDATA: "YES" argument tag.

The second OUTPUT method call overwrites existing data set OCTOBEREVENT_REV4 with data retrieved from OERP. The observations in this newer version of OCTOBEREVENT_REV4 are in order by the values of EVENTDATE, SHIFT, ACTIVITY, and EMPID.

Understanding More about REPLACEDUP in Example 5.10

When SAS creates a hash object, it stores the key items in one location and the data items in another. SAS uses the key items to plan the structure of your hash object so that it can efficiently access the data items. If the call to REPLACEDUP modifies the value of a data item, it does *not* also modify the same-named key item. The entries in your hash object continue to have the same set of key values that it started out with. Even if your hash object was defined as an ordered hash object, the order in which SAS retrieves entries from the hash object does not change.

Note that EMPID in Example 5.10 is both a key item and a data item. When REPLACEDUP executes, it replaces only the EMPID data item. It does *not* alter the value of the EMPID key item. Therefore, when SAS creates the first version of OCTOBEREVENT_REV4 by executing the first OUTPUT method, it retrieves entries from MULT4 in order by the values of the key items that existed when it created MULT4. SAS writes only data items to a data set. Therefore, while the observations are output in order by the key values, they are not in order by the data item values of EMPID. The observations for 724RTQ are now in the same place as the observations for ADHW3A that they replaced.

When SAS creates hash object OERP from OCTOBEREVENT_REV4, its key structure is based on the revised values of EMPID. When SAS creates the newer version of OCTOBEREVENT_REV4 with the second call to the OUTPUT method, it retrieves entries from OERP in order by the revised values of EMPID, as Output 5.14 shows.

Example 5.10 Replacing Data in Specific Sets of Data Items

```
data _null_;
  attrib eventdate format=mmddyy10.
         shift length=$2 label='Event Shift'
         activity length=$20 label='Event Activity'
         empid length=$6 label='Employee ID'
         empname length=$40 label='Employee Name';

  declare hash mult4(dataset: 'octoberevent', multidata: 'yes',
                     ordered: 'yes');
  mult4.definekey('empid');
  mult4.definedata('empid','empname','eventdate','shift','activity');
  mult4.definedone();

  call missing(eventdate,shift,activity,empid,empname);

  rc=mult4.find(key: '02QYJG');
  if rc=0 then do;
    do until (rc ne 0);
      if shift='AM' then activity='Immunization';
      else if shift='PM' then activity='Vitals';
      rcr=mult4.replacedup();
      rc=mult4.find_next();
    end;
  end;

  rc=mult4.find(key: 'ADHW3A');
  if rc=0 then do;
```

```
        do until (rc ne 0);
          if shift='PM' then do;
            empid='724RTQ';
            empname='McCormick, Annemarie L.';
            rcr=mult4.replacedup();
          end;
          rc=mult4.find_next(key: 'ADHW3A');
        end;
      end;

      rc=mult4.output(dataset: 'octoberevent_rev4');

      declare hash oerp(dataset: 'octoberevent_rev4',ordered: 'yes');
      oerp.definekey('eventdate','shift','activity','empid');
      oerp.definedata('empid','empname','eventdate','shift','activity');
      oerp.definedone();

      rc=oerp.output(dataset: 'octoberevent_rev4');

    run;
```

The PROC PRINT report in Output 5.14 includes all of the observations for employees 02QYJG, ADHW3A, and 724RTQ in data set OCTOBEREVENT_REV4. The observations for these three employees are in **bold**. Employee ADHW3A now only has morning shifts the two days. All activities for 02QYJG have been changed. Employee 724RTQ has been assigned the afternoon shifts that were previously assigned to ADHW3A.

Output 5.14 PROC PRINT of OCTOBEREVENT (includes all observations for EMPID="02QYJG", EMPID="ADHW3A", and EMPID="724RTQ")

Obs	empid	empname	eventdate	shift	activity
1	8I4YKY	Brooks, Shelby B.	10/22/2013	AM	Cholesterol
2	**ADHW3A**	**Moore, Allison W.**	**10/22/2013**	**AM**	**Cholesterol**
	. . .				
11	PTBHUP	Thompson, Olivia P.	10/22/2013	AM	Dietary
12	**02QYJG**	**Howard, Rachel Y.**	**10/22/2013**	**AM**	**Immunization**
13	1CV01H	Brown, Lindsey V.	10/22/2013	AM	Immunization
	. . .				
41	**02QYJG**	**Howard, Rachel Y.**	**10/22/2013**	**PM**	**Vitals**
42	14ZN75	Miller, Sierra Q.	10/22/2013	PM	Vitals
43	**724RTQ**	**McCormick, Annemarie L.**	**10/22/2013**	**PM**	**Vitals**
44	7YN5NV	Perez, Paige O.	10/22/2013	PM	Vitals
	. . .				
54	81DJ0Z	Martinez, Kelly I.	10/23/2013	AM	Counseling

Obs	empid	empname	eventdate	shift	activity
55	**ADHW3A**	**Moore, Allison W.**	**10/23/2013**	**AM**	**Counseling**
56	FTZMFI	Rodriguez, Samantha E.	10/23/2013	AM	Counseling
	. . .				
61	K42ODX	Anderson, Rebecca J.	10/23/2013	AM	Dietary
62	**02QYJG**	**Howard, Rachel Y.**	**10/23/2013**	**AM**	**Immunization**
63	67Z8K3	Moore, Vanessa I.	10/23/2013	AM	Immunization
	. . .				
77	WRH1W5	Chou, Alexandria U.	10/23/2013	AM	Vitals
78	**724RTQ**	**McCormick, Annemarie L.**	**10/23/2013**	**PM**	**Cholesterol**
79	7YN5NV	Perez, Paige O.	10/23/2013	PM	Cholesterol
	. . .				

Example 5.10 creates two hash objects sequentially in its processing. Once the DATA step executes the first OUTPUT method call, it no longer needs the first hash object MULT4. It is possible to delete a hash object with the DELETE method or empty a hash object with the CLEAR method and reuse the same hash object during execution of a DATA step. Deleting an unneeded hash object might be important if your computer has memory constraints and the sizes of your hash objects within the DATA step are large. Chapter 6 describes the DELETE and CLEAR methods.

Summarizing Data in Hash Objects That Allow Multiple Sets of Data Items per Key Value

Summary information about the keys in your hash object can be calculated if you specify the SUMINC argument tag on the DECLARE statement that defines the hash object and then use the SUM method to retrieve the summary value. Example 4.11 showed how to use the SUM method and a hash iterator object to sum a variable's value and retrieve this sum for each key value in a hash object. Each time a FIND or REF method executes successfully, SAS adds to the key value's summary the value specified as the object of the SUMINC argument tag.

SAS designed the SUM method to work with hash objects that allow only one set of data items per key value. If you use SUM when your hash object allows multiple sets of data items per key value, SAS retrieves the summary value for the first set of data items for the current key value's even if your code is positioned at a later set of data items for that key value.

When your hash object allows multiple sets of data items per key value, use the SUMDUP method instead of the SUM method. Each time a FIND, FIND_NEXT, or FIND_PREV method executes, SAS adds the value specified as the object of the SUMINC argument to the summary for the current set of data items for a specific key. When you apply the SUMDUP method, SAS retrieves the summary value for the set of data items that your code is positioned at for the specific key value.

The SUMDUP method is limited compared to working with PROC MEANS or PROC FREQ in the situation of multiple sets of data items per key value. This book omits an example of its usage.

Application Example: Summarizing and Sorting a Data Set

Example 5.11 takes advantage of the ORDERED and MULTIDATA argument tags so that a data set can be summarized and then output in a specific order. The hash object allows multiple sets of data items per key value by defining the hash object with the MULTIDATA: "YES" argument tag. SAS retrieves data in descending order of the key values by defining the ORDERED: "D" argument tag. A hash iterator object traverses the hash object.

Example 5.11 processes one data set, FLUSHOTS, which contains the influenza immunization counts for adults at several clinics on several days in October, November, and December. The goal of Example 5.11 is to review the immunization counts and output them in descending order by the number of immunizations given to adults 65 and over, which is stored in variable SHOTS65PLUS. Example 5.11 sums the counts for adults 18 to 64 and for adults 65 and over and saves the value in TOTALSHOTS. It also assigns the day of the week that the clinic was held to variable DAYOFWEEK.

Output 5.15 shows the first 25 of the 65 observations in FLUSHOTS.

Output 5.15 PROC PRINT of FLUSHOTS (first 25 observations)

Obs	clinicsite	clinicdate	shots18_64	shots65plus
1	Downtown Community Center	10/24/2011	73	66
2	Township Hall	10/24/2011	83	53
3	Westside School	10/24/2011	107	114
4	ABC Pharmacy	10/25/2011	114	169
5	Lakeside Corporation	10/25/2011	74	81
6	Lakeside Corporation	10/25/2011	99	69
7	Township Hall	10/25/2011	109	60
8	ABC Pharmacy	10/27/2011	124	172
9	Lakeside Corporation	10/29/2011	74	99
10	Parkview School	10/29/2011	66	82
11	Parkview School	10/29/2011	125	185
12	Parkview School	10/29/2011	122	63
13	Parkview School	10/29/2011	81	80
14	Westside School	10/29/2011	72	71

Obs	clinicsite	clinicdate	shots18_64	shots65plus
15	ABC Pharmacy	11/02/2011	116	136
16	Campus Center	11/02/2011	124	103
17	Campus Center	11/02/2011	120	112
18	Parkview School	11/02/2011	84	103
19	Campus Center	11/03/2011	86	70
20	Campus Center	11/03/2011	125	73
21	Downtown Community Center	11/03/2011	95	65
22	Downtown Community Center	11/03/2011	88	44
23	Lakeside Corporation	11/03/2011	62	50
24	Township Hall	11/03/2011	81	109
25	Westside School	11/03/2011	55	54

The DATA step starts by defining hash object FS and loading into it the observations from FLUSHOTS. The second DECLARE statement defines hash iterator object FSITER and associates it with hash object FS. Since the goal is to order the observations in descending order by the values of SHOTS65PLUS, the DECLARE statement includes the ORDERED: "D" argument tag and the key item is SHOTS65PLUS.

The MULTIDATA: 'YES' argument tag allows multiple sets of data items for a specific value of SHOTS65PLUS in hash object FS. By chance, some clinics might have the same number of shots administered to patients 65 or older. If you do not specify MULTIDATA:'YES', SAS loads into FS only the data from the first occurrence of the SHOTS65PLUS value and it does not warn you that additional sets exist.

The FIRST method call starts the iteration through FS at the first entry in FS, and it returns data for the clinic that has the highest value for variable SHOTS65PLUS. The NEXT method in the DO WHILE loop continues accessing items from FS in descending order of SHOTS65PLUS. The OUTPUT statement writes to data set ORDERED65PLUS the data retrieved from each entry in FS.

Example 5.11 Sorting and Summarizing a Data Set

```
data ordered65plus;
  attrib clinicsite length=$50
         clinicdate length=8 format=mmddyy10.
         dayofweek  length=$9
         shots18_64 label='Adults 18-64 Immunized'
         shots65plus label='Adults 65+ Immunized'
         totalshots  label='Adults Immunized';

  keep clinicsite clinicdate dayofweek shots18_64 shots65plus
totalshots;

  declare hash fs(dataset: 'flushots', ordered: 'd',
                  multidata: 'yes');
```

```
    declare hiter fsiter('fs');
    fs.definekey('shots65plus');
    fs.definedata('shots65plus','clinicdate','clinicsite','shots18_64');
    fs.definedone();
    call missing(clinicsite,clinicdate,shots18_64,shots65plus);

    rc=fsiter.first();
    do while (rc=0);
      dayofweek=left(put(clinicdate,downame9.));
      totalshots=sum(shots18_64,shots65plus);
      output;
      rc=fsiter.next();
    end;
  run;
```

Output 5.16 shows the first 25 observations in output data set ORDERED65PLUS. Note that observations 12, 13, and 14 have identical values for SHOTS65PLUS.

Output 5.16 PROC PRINT of ORDERED65PLUS (first 25 observations)

Obs	clinicsite	clinicdate	dayofweek	shots18_64	shots65plus	totalshots
1	Westside School	11/09/2011	Wednesday	126	189	315
2	Parkview School	10/29/2011	Saturday	125	185	310
3	ABC Pharmacy	10/27/2011	Thursday	124	172	296
4	ABC Pharmacy	10/25/2011	Tuesday	114	169	283
5	Downtown Community Center	11/04/2011	Friday	131	166	297
6	Campus Center	11/11/2011	Friday	136	159	295
7	Westside School	11/11/2011	Friday	106	154	260
8	Westside School	11/04/2011	Friday	99	149	248
9	Township Hall	11/11/2011	Friday	119	148	267
10	Lakeside Corporation	12/01/2011	Thursday	112	139	251
11	ABC Pharmacy	11/02/2011	Wednesday	116	136	252
12	Campus Center	11/09/2011	Wednesday	114	130	244
13	Downtown Community Center	11/09/2011	Wednesday	126	130	256
14	Lakeside Corporation	11/14/2011	Monday	99	130	229
15	ABC Pharmacy	11/11/2011	Friday	117	125	242
16	ABC Pharmacy	11/14/2011	Monday	96	120	216
17	Parkview School	12/02/2011	Friday	111	119	230
18	Downtown Community Center	12/02/2011	Friday	126	118	244
19	Westside School	10/24/2011	Monday	107	114	221
20	Campus Center	11/02/2011	Wednesday	120	112	232
21	Township Hall	11/03/2011	Thursday	81	109	190

Obs	clinicsite	clinicdate	dayofweek	shots18_64	shots65plus	totalshots
22	Campus Center	11/11/2011	Friday	91	109	200
23	Campus Center	11/02/2011	Wednesday	124	103	227
24	Parkview School	11/02/2011	Wednesday	84	103	187
25	ABC Pharmacy	11/05/2011	Saturday	75	102	177

Application Example: Creating Data Sets Based on a Series of Observations

Example 5.12 uses two hash objects to classify the observations in a data set for output to specific data sets. The example takes advantage of the default definition of a hash object that allows only one set of data items per key value. When the data contains multiple sets of data items per key value and the hash object definition specifies the DUPLICATE: REPLACE argument tag, SAS retains in the hash object the last set of data items for the key value. Example 5.12 uses information from the last set of data items for a key value to classify all of the observations in the example's data set, which it loads into the second hash object. The second hash object allows multiple sets of data items per key value.

A hash iterator object traverses the first hash object. The code reads each key value in the first hash object and its associated last set of data items. SAS then retrieves from the second hash object all sets of data items for each key value. The value of a specific data item returned from the first hash object classifies the data retrieved from the second hash object for each key value. SAS outputs all of the sets of data items for a key value to a data set whose name reflects the key value's data item retrieved from the first hash object.

Example 5.12 processes data set STUDYAPPTS. This data set contains height, weight, and blood pressure measurements for 24 subjects in a clinical study. The measurements were taken at five different intervals during the study: the start of the study (baseline); 30 days; 60 days; 120 days; and 180 days. Each observation in STUDYAPPTS contains the data for one subject's measurements at one of the five intervals.

Assume that the subjects started the study at different times and that the study has not yet finished. Therefore, a subject may have between one and five observations in STUDYAPPTS. All subjects in STUDYAPPTS have at least one measurement in addition to the baseline measurement.

The goal of Example 5.12 is to create five data sets, one for each of the five measurement times. Example 5.12 outputs all observations for a subject to the data set whose name corresponds to the last appointment that was recorded for the subject.

Output 5.17 shows data for the first five subjects in STUDYAPPTS. Reviewing the data for these five subjects, their cohorts are as follows.

- Study ID AFF115 is in the two-month appointment cohort.

- Study IDs BBR123 and DDL118 are in the six-month appointment cohort.
- Study ID CDF108 is in the four-month appointment cohort.
- Study ID DHZ122 is in the one-month appointment cohort.

Output 5.17 PROC PRINT of Data for Five Subjects in STUDYAPPTS

Obs	studyid	apptdate	appttype	weight	systol	diast
1	AFF115	03/07/2013	Baseline	175	153	88
2	AFF115	04/05/2013	One Month	174	151	84
3	AFF115	06/28/2013	Two Month	169	155	82
4	BBR123	03/08/2013	Baseline	163	155	93
5	BBR123	04/11/2013	One Month	158	162	91
6	BBR123	07/03/2013	Two Month	163	158	93
7	BBR123	11/01/2013	Four Month	168	161	93
8	BBR123	05/12/2014	Six Month	160	147	87
9	CDF108	02/04/2013	Baseline	244	143	89
10	CDF108	03/15/2013	One Month	252	147	91
11	CDF108	05/20/2013	Two Month	249	152	92
12	CDF108	10/11/2013	Four Month	256	152	94
13	DDL118	02/19/2013	Baseline	187	162	95
14	DDL118	03/28/2013	One Month	185	166	95
15	DDL118	06/07/2013	Two Month	181	167	93
16	DDL118	10/24/2013	Four Month	174	167	91
17	DDL118	05/22/2014	Six Month	170	157	84
18	DHZ122	03/08/2013	Baseline	138	142	83
19	DHZ122	04/18/2013	One Month	141	137	81

The DATA statement names the five output data sets, one for each of the five times that measurements were made.

Hash object LASTAPPT identifies the last appointment for each subject. The subject's ID, STUDYID, is the key item in LASTAPPT. When you use the DUPLICATE: "REPLACE" argument tag, SAS loads each measurement time for each subject into hash object LASTAPPT, one measurement time after the previous, and it overwrites the previous measurement time when it processes the second and subsequent measurements for a subject. The last measurement time remains in the hash object.

Data set STUDYAPPTS must be sorted by appointment date before the DATA step executes so that SAS keeps the latest date and associated appointment type for each subject in LASTAPPT. The RENAME= option applied to STUDYAPPTS renames APPTTYPE to LASTAPPTTYPE so that this item's value does not get overwritten or replaced by the retrieval of the APPTTYPE data item from the second hash object, ALLAPPTS.

SAS loads all of the observations in STUDYAPPTS into hash object ALLAPPTS. This hash object is also keyed by STUDYID. The MULTIDATA: "YES" argument tag is required so that all observations for each subject are in ALLAPPTS.

Hash iterator object LASTITER traverses hash object LASTAPPT. The DATA step starts at the first entry in LASTAPPT. The FIND method then searches for the first entry in ALLAPPTS for the value of STUDYID returned from LASTAPPT. The DO UNTIL loop and the FIND_NEXT method find all of the remaining entries for the current value of STUDYID.

Hash object LASTAPPT also returns data item LASTAPPTTYPE. The IF-THEN-ELSE statements in the DO UNTIL loop write a subject's observations to the data set specified by the value of LASTAPPTYPE.

Example 5.12 Creating Data Sets Based on a Series of Observations

```
data baseline onemonth twomonth fourmonth sixmonth;
  attrib studyid   length=$6
         appttype  length=$15
         lastappttype length=$15
         apptdate  length=8 format=mmddyy10.
         weight    length=8
         systol    length=8
         diast     length=8;

  declare hash lastappt(dataset: 'studyappts
                        (rename=(appttype=lastappttype))',
                        duplicate: 'replace',
                        ordered: 'yes');
  declare hiter lastiter('lastappt');
  lastappt.definekey('studyid');
  lastappt.definedata('studyid','lastappttype');
  lastappt.definedone();

  declare hash allappts(dataset: 'studyappts', multidata: 'yes',
                        ordered: 'yes');
  allappts.definekey('studyid');
  allappts.definedata('studyid','appttype','apptdate','weight',
                      'systol', 'diast');
  allappts.definedone();

  call missing(studyid,appttype,lastappttype,apptdate,weight,systol,diast);
  drop rc rciter;

  rciter=lastiter.first();
  do until (rciter ne 0);
    rc=allappts.find();
```

```
      do until (rc ne 0);
        if lastappttype='Baseline' then output baseline;
        else if lastappttype='One Month' then output onemonth;
        else if lastappttype='Two Month' then output twomonth;
        else if lastappttype='Four Month' then output fourmonth;
        else if lastappttype='Six Month' then output sixmonth;
        rc=allappts.find_next();
      end;
      rciter=lastiter.next();
    end;
  run;
```

Output 5.18 lists the observations in data set ONEMONTH.

Output 5.18 PROC PRINT of ONEMONTH

Obs	studyid	appttype	lastappttype	apptdate	weight	systol	diast
1	DHZ122	Baseline	One Month	03/08/2013	138	142	83
2	DHZ122	One Month	One Month	04/18/2013	141	137	81
3	MBO120	Baseline	One Month	02/11/2013	203	138	85
4	MBO120	One Month	One Month	03/21/2013	207	144	81
5	NDM112	Baseline	One Month	02/13/2013	161	128	88
6	NDM112	One Month	One Month	03/18/2013	165	129	87

Output 5.19 lists the observations in data set FOURMONTH.

Output 5.19 PROC PRINT of FOURMONTH

Obs	studyid	appttype	lastappttype	apptdate	weight	systol	diast
1	CDF108	Baseline	Four Month	02/04/2013	244	143	89
2	CDF108	One Month	Four Month	03/15/2013	252	147	91
3	CDF108	Two Month	Four Month	05/20/2013	249	152	92
4	CDF108	Four Month	Four Month	10/11/2013	256	152	94
5	LIM109	Baseline	Four Month	02/12/2013	173	137	76
6	LIM109	One Month	Four Month	03/22/2013	181	130	77
7	LIM109	Two Month	Four Month	06/10/2013	185	128	77
8	LIM109	Four Month	Four Month	10/18/2013	183	129	77
9	LOE121	Baseline	Four Month	03/15/2013	255	149	82
10	LOE121	One Month	Four Month	04/22/2013	259	146	82
11	LOE121	Two Month	Four Month	07/10/2013	269	150	83
12	LOE121	Four Month	Four Month	11/12/2013	269	147	78
13	MIJ104	Baseline	Four Month	02/08/2013	205	135	90
14	MIJ104	One Month	Four Month	03/22/2013	212	132	90

Obs	studyid	appttype	lastappttype	apptdate	weight	systol	diast
15	MIJ104	Two Month	Four Month	06/10/2013	218	134	85
16	MIJ104	Four Month	Four Month	10/28/2013	220	135	84
17	NHC105	Baseline	Four Month	03/15/2013	222	148	82
18	NHC105	One Month	Four Month	04/15/2013	232	154	85
19	NHC105	Four Month	Four Month	08/22/2013	218	154	84
20	REA102	Baseline	Four Month	03/07/2013	154	173	85
21	REA102	One Month	Four Month	04/16/2013	154	178	89
22	REA102	Two Month	Four Month	06/27/2013	148	170	85
23	REA102	Four Month	Four Month	11/08/2013	142	159	83

Application Example: Creating a Data Set That Contains All Combinations of Specific Variables When the Number of Combinations Is Large

Example 5.13 shows that a hash object in a DATA step can be a quick and easy way to find all the combinations of specific variables in a data set. It loads a data set into a hash object, and it takes advantage of the default processing of a hash object where multiple sets of data items are not allowed.

The variables whose combinations you want to find are specified as both key items and data items in the hash object. SAS loads only the unique combinations of the values of the key items into the hash object. Then it applies the OUTPUT method to write the contents of this hash object to a data set that contains all of the unique combinations of the specific variables.

A reason you might want to use this solution is if the number of combinations in your data set is very large. A typical way of finding all the combinations of specific variables is to use PROC FREQ. However, depending on memory resources, the number of combinations in your data set might be too large for PROC FREQ to analyze. SAS might write either of the following error messages in that situation:

```
ERROR: The requested table is too large to process.

ERROR: The SAS System stopped processing this step because of
insufficient memory.
```

A data set that contains all of the unique combinations of specific variables can serve as the CLASSDATA= data set for PROC MEANS or PROC TABULATE. You specify a CLASSDATA= data set when you want PROC MEANS or PROC TABULATE to produce output that contains all the combinations in that data set. Even when an analysis variable has no data for a combination, PROC MEANS or PROC TABULATE still inserts an entry for that combination in the table or data set that the procedure produces when you specify a CLASSDATA= data set.

The goal of Example 5.13 is to find all the combinations of variables PTID, PROVIDER_TYPE, and CLAIMDATE in the CLAIMS2013 data set. This data set contains several thousand observations of medical claims data. Each observation contains data for one medical claim for one patient on one date. Assume that your computer does not have sufficient memory to execute the following PROC FREQ TABLES statement:

```
tables ptid*claimdate*provider_type / noprint out=claimcombos;
```

While you could use BY-variable processing in a DATA step to find the set of combinations, this method requires that you sort or index your input data set by the variables on the BY statement.

Output 5.20 shows the first 25 observations in CLAIMS2013.

Output 5.20 PROC PRINT of CLAIMS2013 (first 25 observations)

Obs	ptid	claimdate	provider_id	charge	provider_type
1	4DGNWU3Z	04/12/2013	OPSURGERY039	$7,902.00	OPSURGERY
2	4DGNWU3Z	02/07/2013	PHYSTHERAPY071	$330.40	PHYSTHERAPY
3	XZRQTXGU	04/17/2013	OPRADIOLOGY069	$1,517.60	OPRADIOLOGY
4	JT1Z17GE	08/22/2013	CLINIC016	$53.00	CLINIC
5	THSALN2J	09/25/2013	OPRADIOLOGY070	$450.20	OPRADIOLOGY
6	THSALN2J	11/11/2013	CLINIC014	$229.80	CLINIC
7	THSALN2J	08/15/2013	CLINIC010	$102.20	CLINIC
8	THSALN2J	09/09/2013	LAB049	$147.40	LAB
9	DWJSEGFQ	07/31/2013	PHYSTHERAPY074	$381.40	PHYSTHERAPY
10	DWJSEGFQ	10/04/2013	CLINIC023	$65.80	CLINIC
11	L2PAT59T	10/16/2013	OPSURGERY044	$3,373.80	OPSURGERY
12	GE356WUG	10/07/2013	OPSURGERY035	$7,128.20	OPSURGERY
13	GE356WUG	07/24/2013	CLINIC005	$104.60	CLINIC
14	96O0JS7E	10/16/2013	CLINIC003	$280.40	CLINIC
15	96O0JS7E	04/08/2013	OPSURGERY036	$7,243.60	OPSURGERY
16	QMPH9WU3	11/04/2013	OPSURGERY041	$1,428.60	OPSURGERY
17	QMPH9WU3	07/03/2013	OPRADIOLOGY070	$3,180.20	OPRADIOLOGY
18	AD0ONT9C	03/21/2013	CLINIC012	$62.00	CLINIC
19	ZLEYIOE7	01/18/2013	PHYSTHERAPY074	$209.40	PHYSTHERAPY
20	ZLEYIOE7	08/20/2013	CLINIC002	$105.20	CLINIC
21	ZLEYIOE7	09/12/2013	CLINIC012	$168.60	CLINIC
22	ZLEYIOE7	04/09/2013	OPSURGERY035	$5,686.20	OPSURGERY

Obs	ptid	claimdate	provider_id	charge	provider_type
23	ZLEYIOE7	05/23/2013	LAB059	$137.20	LAB
24	8WQJ7DRM	09/05/2013	LAB049	$421.40	LAB
25	8WQJ7DRM	08/01/2013	CLINIC025	$88.80	CLINIC

The DATA step in Example 5.13 is simple. Its first action creates hash object COMBO from data set CLAIMS2013. The variables whose combinations you want to find are specified as key items and as data items. The MULTIDATA: argument tag is *not* specified. Therefore, SAS does not allow multiple sets of data items per combination of the key item values. The ORDERED: 'YES' argument tag is specified so that the OUTPUT method creates a data set that is in order by the combinations of the key item values, but this specification is not required.

The last statement in Example 5.13 step applies the OUTPUT method to create data set CLAIMCOMBOS from hash object COMBO.

Example 5.13 Creating a Data Set That Contains All Combinations of Specific Variables

```
data _null_;
  attrib ptid length=$8 label='Patient ID'
         claimdate length=8 format=mmddyy10. label='Claim Date'
         provider_type length=$11 label='Provider Type';

  declare hash combo(dataset: 'claims2013',ordered: 'yes');
  combo.definekey('ptid','claimdate','provider_type');
  combo.definedata('ptid','claimdate','provider_type');
  combo.definedone();

  call missing(ptid,claimdate,provider_type);

  rc=combo.output(dataset: 'claimcombos');
run;
```

Output 5.21 shows the first 15 observations in data set CLAIMCOMBOS.

Output 5.21 PROC PRINT of CLAIMCOMBOS (first 15 observations)

Obs	ptid	claimdate	provider_type
1	000D03OP	03/08/2013	LAB
2	000M55K7	10/10/2013	CLINIC
3	000Q7RE7	07/18/2013	CLINIC
4	0010YCCP	08/08/2013	CLINIC
5	001QHXJC	04/19/2013	CLINIC
6	002ETJ20	11/13/2013	LAB
7	0030W4JD	05/28/2013	PHYSTHERAPY
8	0030W4JD	10/02/2013	OPSURGERY

Obs	ptid	claimdate	provider_type
9	0033XMC8	09/11/2013	CLINIC
10	0034S0O5	04/15/2013	LAB
11	0034S0O5	09/18/2013	OPSURGERY
12	003PN3TO	03/05/2013	CLINIC
13	004EAAGF	08/22/2013	PHYSTHERAPY
14	004FZD21	03/01/2013	LAB
15	004FZD21	08/05/2013	OPRADIOLOGY

Application Example: Linking Hierarchically Related Data Using a Hash Object That Allows Multiple Sets of Data Items per Key Value

Example 5.14 shows how a hash object and hash iterator object can link two data sets that are hierarchically related and how they can summarize the data from the data set lower in the hierarchy. This example produces the same data set that Example 3.13 does. The difference between the two examples is that Example 5.14 uses a hash object that allows multiple sets of data items per key value.

The data used in Example 5.14 is from a survey where information was gathered about households and the people living in them. Data set HH stores the general information about the household. Data set PERSONS stores data about each person living in the households. Variable HHID uniquely identifies each household, and this variable is common to both data sets. The two data sets are hierarchically related. Each household can have one or more people living in the household. Observations in PERSONS are uniquely identified by two variables, the household ID variable HHID and the person ID variable PERSONID.

The goal of Example 5.14 is to create a data set with one observation per household. This observation should contain the general data recorded for each household in HH and a summary of the person information in PERSONS for each household. The DATA step uses a hash object and hash iterator object to process the person records so that it can determine four statistics for each household:

- total number of people living in the household
- number of people in each of three age groups
- highest level of education in the household
- total income in the household

When the households were surveyed, the persons interviewed were assigned a sequential ID value starting with 1. All households have at least one member.

Output 5.22 shows the contents of HH. Data set HH contains data for 10 households.

Output 5.22 PROC PRINT of HH

Obs	hhid	tract	surveydate	hhtype
1	HH01	CS	07/09/2012	Owner
2	HH02	CN	03/11/2012	Renter
3	HH03	CS	05/20/2012	Owner
4	HH04	SW	01/12/2012	Owner
5	HH05	NE	10/17/2012	Renter
6	HH06	NE	05/15/2012	Owner
7	HH07	SW	02/02/2012	Owner
8	HH08	NE	04/09/2012	Renter
9	HH09	CE	11/01/2012	Owner
10	HH10	CN	03/31/2012	Owner

Data set PERSONS contains at least one observation for every household in HH. Output 5.23 shows the contents of PERSONS.

Output 5.23 PROC PRINT of PERSONS

Obs	hhid	personid	age	gender	income	educlevel
1	HH01	P01	68	M	$52,000	12
2	HH01	P02	68	F	$23,000	12
3	HH02	P01	42	M	$168,100	22
4	HH03	P01	79	F	$38,000	10
5	HH04	P01	32	F	$56,000	16
6	HH04	P02	31	M	$72,000	18
7	HH04	P03	5	F	$0	0
8	HH04	P04	2	F	$0	0
9	HH05	P01	26	M	$89,000	22
10	HH06	P01	56	M	$123,000	18
11	HH06	P02	48	F	$139,300	18
12	HH06	P03	17	F	$5,000	11
13	HH07	P01	48	M	$90,120	16
14	HH07	P02	50	F	$78,000	18
15	HH08	P01	59	F	$55,500	16
16	HH09	P01	32	F	$48,900	14
17	HH09	P02	10	M	$0	5
18	HH10	P01	47	F	$78,000	16
19	HH10	P02	22	F	$32,000	16
20	HH10	P03	19	M	$20,000	12

Obs	hhid	personid	age	gender	income	educlevel
21	HH10	P04	14	M	$0	8
22	HH10	P05	12	F	$0	6
23	HH10	P06	9	F	$0	4

Example 5.14 starts by defining hash object P. It loads all observations from PERSONS into P. The entries in hash object P are keyed and SAS retrieves data from P in order by the values of HHID. Since there can be more than one person per household, the DECLARE statement includes the MULTIDATA: "YES" argument tag.

The SET statement reads the observations in HH one at a time. Each iteration of the DATA step initializes the six summary variables to 0. Variables NPERSONS, NPLT18, NP18_64, and NP65PLUS are tallies of the total number of people in the household and the number in each of three age groups: less than 18; 18–64; and greater than 64. Variable HHINCOME is a tally of the total income earned by all members of the household. Variable HIGHESTED is the maximum value of EDUCLEVEL in the household.

The FIND method finds the first occurrence in P for the current value of HHID. Statements in the DO UNTIL loop summarize the person information for the household. The DO UNTIL loop executes the FIND_NEXT method to read data for each person in the household. The loop ends when SAS has finished reading all person data for a household.

Example 5.14 Linking Hierarchically Related Data Using a Hash Object That Allows Multiple Sets of Data Items per Key Value

```
data hhsummary;
  attrib hhid length=$4
         tract length=$4
         surveydate format=mmddyy10.
         hhtype length=$10
         personid length=$4
         age length=3
         gender length=$1
         income length=8 format=dollar12.
         educlevel length=3
         npersons length=3
         nplt18 length=3
         np18_64 length=3
         np65plus length=3
         highested length=3
         hhincome  length=8 format=dollar12.;

  if _n_=1 then do;
    declare hash p(dataset: 'persons',ordered: 'yes', multidata: 'yes');
    p.definekey('hhid');
    p.definedata('personid','age','gender','income','educlevel');
    p.definedone();
```

```
    call missing(personid,age,gender,income,educlevel);
  end;

  keep hhid tract surveydate hhtype npersons nplt18 np18_64 np65plus
       highested hhincome;
  array zeroes{*} npersons nplt18 np18_64 np65plus hhincome highested;

  set hh;

  do i=1 to dim(zeroes);
    zeroes{i}=0;
  end;

  rc=p.find();
  do until(rc ne 0);
    npersons+1;
    if age lt 18 then nplt18+1;
    else if 18 le age le 64 then np18_64+1;
    else if age ge 65 then np65plus+1;
    hhincome+income;
    if educlevel gt highested then highested=educlevel;
    rc=p.find_next();
  end;
run;
```

Output 5.24 lists the observations in HHSUMMARY. The output is identical to that shown in Output 3.34.

Output 5.24 PROC PRINT of HHSUMMARY

Obs	hhid	tract	surveydate	hhtype	npersons	nplt18	np18_64	np65plus	highested	hhincome
1	HH01	CS	07/09/2012	Owner	2	0	0	2	12	$75,000
2	HH02	CN	03/11/2012	Renter	1	0	1	0	22	$168,100
3	HH03	CS	05/20/2012	Owner	1	0	0	1	10	$38,000
4	HH04	SW	01/12/2012	Owner	4	2	2	0	18	$128,000
5	HH05	NE	10/17/2012	Renter	1	0	1	0	22	$89,000
6	HH06	NE	05/15/2012	Owner	3	1	2	0	18	$267,300
7	HH07	SW	02/02/2012	Owner	2	0	2	0	18	$168,120
8	HH08	NE	04/09/2012	Renter	1	0	1	0	16	$55,500
9	HH09	CE	11/01/2012	Owner	2	1	1	0	14	$48,900
10	HH10	CN	03/31/2012	Owner	6	3	3	0	16	$130,000

Chapter 6: Managing Hash Objects

This chapter focuses on the hash methods and argument tags that can help you manage creation and deletion of hash objects. These methods allow you to dynamically create, clear, delete, and compare hash objects during execution of a DATA step. When your applications are complex and have the potential to consume too much memory or conflict with other processes, you can adjust how SAS allocates memory when it defines the hash object's structure. An attribute can tell you how much memory an entry takes so that you can better determine how to make this adjustment. Another attribute can track the number of items in a hash object.

The examples in this chapter show a few ways that hash objects can be managed.

Creating, Deleting, and Clearing Hash Objects During Execution of a DATA Step

All of the examples in the previous chapters created hash objects during execution of a DATA step. Most of the DATA steps in these examples started with a block of code that executed once on the first iteration of the DATA step and created the hash objects that the DATA steps used. When these DATA steps ended, SAS automatically deleted all hash objects it created during execution of the DATA step, and it cleared the memory that it used to store the hash objects.

You are not limited to creating hash objects once during execution of a DATA step. You can issue the set of statements that define hash objects as many times as needed in your DATA step. Further, if memory constraints exist either because your hash objects are large or because your computer has limited memory resources, you can delete hash objects during execution of a DATA step by calling the DELETE method. This action frees up memory before your DATA step ends.

You can also keep a hash object and clear its contents by calling the CLEAR method. This technique allows you to reuse a hash object. While you could call the REMOVE method the same number of times as there are entries in the hash object, the CLEAR method more efficiently empties the hash object for you since you need to execute only one method call.

You cannot overwrite an existing hash object by issuing a DECLARE statement that defines a new hash object with the same name. You must first delete the hash object before you redefine it.

The syntax for the DELETE and CLEAR methods follows. Neither method has any argument tags.

```
rc=object.CLEAR();

rc=object.DELETE();
```

Example 6.1 calls the DELETE method. Example 6.3 calls the CLEAR method and reuses an existing hash object.

The calls to the DELETE method in Example 6.1 remove a hash object before the DATA step creates another hash object. On each iteration of the DO loop, SAS creates and deletes hash objects C, C2, and O.

Example 6.1 processes two large data sets, CLAIMS2012 and CLAIMS2013. Both data sets have the same variables and structure. Each observation in these data sets contains data for one medical claim for one patient during the year specified in the data set's name. Output 6.1 shows the first 25 observations in CLAIMS2012.

Output 6.1 PROC PRINT of CLAIMS2012 (first 25 observations)

Obs	ptid	claimdate	provider_id	charge	provider_type
1	NG67JO5E	02/29/2012	CLINIC001	$91.20	CLINIC
2	NG67JO5E	01/24/2012	OPSURGERY046	$4,612.80	OPSURGERY
3	NG67JO5E	03/19/2012	LAB049	$255.40	LAB
4	6O1BUGQG	06/05/2012	LAB055	$172.20	LAB
5	6O1BUGQG	08/30/2012	CLINIC025	$335.20	CLINIC
6	069BEHOM	03/05/2012	CLINIC005	$70.20	CLINIC
7	EQYJRYH1	05/09/2012	CLINIC002	$72.80	CLINIC
8	PADH3JF9	07/16/2012	CLINIC025	$155.80	CLINIC
9	PADH3JF9	02/08/2012	OPSURGERY035	$3,622.00	OPSURGERY
10	PADH3JF9	05/24/2012	CLINIC027	$132.00	CLINIC
11	PADH3JF9	03/05/2012	PHYSTHERAPY075	$185.40	PHYSTHERAPY

Obs	ptid	claimdate	provider_id	charge	provider_type
12	1LRHHEIR	10/02/2012	LAB062	$248.60	LAB
13	1LRHHEIR	03/01/2012	LAB049	$47.00	LAB
14	K7AYGLSM	08/20/2012	LAB062	$386.20	LAB
15	K7AYGLSM	02/23/2012	OPRADIOLOGY066	$1,231.00	OPRADIOLOGY
16	K7AYGLSM	06/07/2012	CLINIC033	$64.00	CLINIC
17	K7AYGLSM	11/13/2012	OPSURGERY041	$6,629.00	OPSURGERY
18	O5NGJEIS	01/31/2012	CLINIC034	$166.60	CLINIC
19	O5NGJEIS	04/03/2012	LAB050	$246.00	LAB
20	OKCKQ0MQ	09/07/2012	OPRADIOLOGY068	$1,794.20	OPRADIOLOGY
21	7RZS72VC	11/13/2012	LAB054	$229.60	LAB
22	7RZS72VC	03/16/2012	LAB053	$59.40	LAB
23	AIMXRGOM	05/04/2012	CLINIC025	$130.80	CLINIC
24	AIMXRGOM	09/26/2012	LAB053	$337.60	LAB
25	I00GNLTI	05/29/2012	CLINIC027	$250.20	CLINIC

The goal of Example 6.1 is to find the maximum charge (CHARGE) for each patient in the year and then create a data set with these observations that is in order by the provider ID (PROVIDER_ID) and patient ID (PTID). Alternatively, these steps can be accomplished with PROC SORT and PROC MEANS, and a macro program could be written that repeats the same PROC steps for each year. A program with these multiple steps may be more efficient and have less I/O than a DATA step and hash object solution. However, Example 6.1 shows that a DATA step that uses hash objects can easily consolidate the processing. You will need to evaluate your application to determine the most efficient solution.

The two assignment statements at the beginning of the DO loop create variables that identify the input data set and the output data set.

The first action in the DO loop creates hash object C from the data set whose name is the current value of variable DSCLAIMS. It defines C so that SAS retrieves data from C in descending order by the values of key items PTID and CHARGE. This ensures that SAS loads the maximum charge for each patient into the hash object. The argument tag MULTIDATA: "YES" allows for multiple sets of data items per key value. Note that if you omitted the MULTIDATA: "YES" argument tag and kept the ORDERED: "DESCENDING" argument tag, SAS would load into C the first observation for each key value in the data set. SAS does *not* order the key values in the entire data set *before* it enters the set of data items when the default setting of MULTIDATA: "NO" is in effect.

The first OUTPUT method call creates a data set whose name is the current value of variable DSOUT. The number of observations in this data set is the same as the number of observations in the year's claims input data set. To conserve memory resources, the DELETE method removes hash object C before SAS creates the next hash object.

The second action in the DO loop saves only the maximum charge for each patient. SAS loads the data set whose name is the current value of variable DSOUT into hash object C2. The MULTIDATA: "YES" argument tag is omitted. With PTID as the key, SAS loads the first occurrence for PTID into C2. The order of the observations in the data set named by variable DSOUT is in descending order by values of PTID and CHARGE because of the action of loading hash object C at the beginning of the DATA step. Therefore, the maximum value of CHARGE for each PTID is stored in C2.

The second OUTPUT method call overwrites the data set whose name is the current value of variable DSOUT with the entries in C2. The number of observations in this data set is equal to the number of unique PTID values in the year's claims data set. The second DELETE method removes hash object C2 before SAS creates the third hash object.

The third action in the DO loop arranges the selected claims data in order by PROVIDER_ID and PTID. SAS loads the observations from the data set produced by the second OUTPUT method call into hash object O. The ORDERED: "YES" argument tag causes the entries to be retrieved in order by the values of PROVIDER_ID and PTID.

The last OUTPUT method call overwrites the data set whose name is the current value of variable DSOUT. The number of observations in this final data set is equal to the number of unique PTID values in the year's claims data set. The observations are in order by PROVIDER_ID and PTID. The RENAME data set option is added to the current value of DSOUT so that SAS changes the name of variable CHARGE to MAXCHARGE_*yyyy* where *yyyy* is either 2012 or 2013.

Example 6.1 Creating and Deleting Hash Objects in the Same DATA Step

```
data _null_;
  attrib ptid length=$8 label='Patient ID'
         claimdate length=8 format=mmddyy10. label='Claim Date'
         provider_id length=$15 label='Provider ID'
         charge length=8 format=dollar10.2 label='Charge'
         dsout length=$100;

  call missing(ptid,claimdate,provider_id,charge,provider_type);

  do claimyear=2012, 2013;
    dsclaims=cats('claims',put(claimyear,4.));
    dsout=cats('maxclaims',put(claimyear,4.),'_sorted');

    declare hash c(dataset: dsclaims, ordered: 'descending',
                   multidata: 'yes');
    c.definekey('ptid','charge');
    c.definedata('ptid','claimdate','charge','provider_id');
    c.definedone();
    rc=c.output(dataset: dsout );
    rc=c.delete();
```

```
      declare hash c2(dataset: dsout, ordered: 'yes');
      c2.definekey('ptid');
      c2.definedata('ptid','claimdate','charge','provider_id');
      c2.definedone();
      rc=c2.output(dataset: dsout );
      rc=c2.delete();

      declare hash o(dataset: dsout, ordered: 'yes',multidata: 'yes');
      o.definekey('provider_id','ptid');
      o.definedata('ptid','claimdate','charge','provider_id');
      o.definedone();

      dsout=cats(dsout,'(rename=(charge=maxcharge_',put(claimyear,4.),')');

      rc=o.output(dataset: dsout );
      rc=o.delete();
    end;
  run;
```

Output 6.2 shows the first 25 observations in data set MAXCLAIMS2012_SORTED.

Output 6.2 PROC PRINT of MAXCLAIMS2012_SORTED (first 25 observations)

Obs	ptid	claimdate	maxcharge_2012	provider_id
1	03E622ZN	02/09/2012	$119.40	CLINIC001
2	040N6YPW	11/21/2012	$151.60	CLINIC001
3	045URYHS	04/16/2012	$131.60	CLINIC001
4	05KCY3H9	06/27/2012	$167.80	CLINIC001
5	05NPT87K	11/01/2012	$113.60	CLINIC001
6	09IUNWGU	04/13/2012	$165.80	CLINIC001
7	0BJND9HC	05/16/2012	$128.00	CLINIC001
8	0CKIS7IL	02/29/2012	$214.20	CLINIC001
9	0CSK08O6	01/26/2012	$114.00	CLINIC001
10	0EU5N67C	11/02/2012	$181.60	CLINIC001
11	0FM0SG5U	07/26/2012	$68.80	CLINIC001
12	0FTZD3AQ	07/09/2012	$186.40	CLINIC001
13	0FWN9I6H	11/14/2012	$66.80	CLINIC001
14	0GXS5G1F	01/19/2012	$74.00	CLINIC001
15	0HXEK7UY	06/20/2012	$124.80	CLINIC001
16	0L9Q5V28	10/24/2012	$127.80	CLINIC001
17	0NM270XI	11/26/2012	$63.60	CLINIC001
18	0QMZSPD0	09/27/2012	$122.60	CLINIC001
19	0SR2KBMS	04/13/2012	$136.40	CLINIC001

Obs	ptid	claimdate	maxcharge_2012	provider_id
20	0T6O9BZM	03/02/2012	$128.40	CLINIC001
21	0UDMUHD3	01/27/2012	$123.80	CLINIC001
22	0UU27MSN	01/09/2012	$138.80	CLINIC001
23	0V0CCRJ4	04/27/2012	$305.40	CLINIC001
24	10BKUKT4	08/28/2012	$172.80	CLINIC001
25	11HQS6SE	01/11/2012	$70.40	CLINIC001

Determining the Number of Items in a Hash Object

When your DATA step application needs to use the number of items in a hash object, you can reference the NUM_ITEMS attribute. This feature applied to a hash object returns the number of items in that hash object. When your hash object does not allow multiple sets of data items per key value, NUM_ITEMS returns the number of unique key values. When your hash object allows multiple sets of data items per key value, NUM_ITEMS returns the total number of rows in the hash object.

The syntax of the NUM_ITEMS attribute follows. Note that you write an attribute slightly differently than you write a method. An attribute does not terminate with a set of parentheses nor does it have any argument tags or options.

```
variable_name=object.NUM_ITEMS;
```

Example 6.2 illustrates NUM_ITEMS with data set BPMEASURES, which was used in Chapter 5. This data set contains seven blood pressure measurements for one patient over a 4-day period. One measurement was recorded on May 13, three on May 14, two on May 15, and one on May 16. BPTIME records the time the measurement was taken. Variable PTID identifies the patient. Output 6.3 shows the contents of BPMEASURES.

Output 6.3 PROC PRINT of BPMEASURES

Obs	ptid	bpdate	bptime	systol	diast
1	AU81750Y	05/13/2013	10:15 AM	131	79
2	AU81750Y	05/14/2013	8:30 AM	125	80
3	AU81750Y	05/14/2013	3:45 PM	141	83
4	AU81750Y	05/14/2013	7:10 PM	132	80
5	AU81750Y	05/15/2013	9:40 AM	125	73
6	AU81750Y	05/15/2013	1:45 PM	133	85
7	AU81750Y	05/16/2013	11:20 AM	128	78

Example 6.2 creates two hash objects, BPUNIQUE and BPMULT. BPUNIQUE does not allow multiple data items value while BPMULT does. The NUM_ITEMS attribute is applied to each of the hash objects. PUT statements write the values NUM_ITEMS returns in the SAS log.

Example 6.2 Reviewing the Usage of NUM_ITEMS

```
data _null_;
  attrib ptid length=$8
         bpdate length=8 format=mmddyy10.
         bptime length=8 format=timeampm8.
         systol length=8 label='Systolic BP'
         diast  length=8 label='Diastolic BP';

  declare hash bpunique(dataset: 'bpmeasures',ordered: 'yes');
  bpunique.definekey('ptid','bpdate');
  bpunique.definedata('ptid','bpdate','bptime','systol','diast');
  bpunique.definedone();

  declare hash bpmult(dataset: 'bpmeasures',ordered: 'yes',multidata: 'yes');
  bpmult.definekey('ptid','bpdate');
  bpmult.definedata('ptid','bpdate','bptime','systol','diast');
  bpmult.definedone();

  call missing(ptid,bpdate,bptime,systol,diast);

  nbpunique=bpunique.num_items;
  nbpmult=bpmult.num_items;
  put '**Number of items in BPUNIQUE=' nbpunique;
  put '**Number of items in BPMULT=' nbpmult;
run;
```

Since measurements were taken on four days and BPDATE is a key item in BPUNIQUE where multiple sets of data items are not allowed, the number of entries in BPUNIQUE is 4.

The DATA step loads data from all seven observations in BPMEASURES into BPMULT. Therefore, NUM_ITEMS returns a value of 7 when applied to BPMULT.

The SAS log for Example 6.2 follows.

```
NOTE: There were 7 observations read from the data set WORK.BPMEASURES.
NOTE: There were 7 observations read from the data set WORK.BPMEASURES.
**Number of items in BPUNIQUE=4
**Number of items in BPMULT=7
```

Application Example: Creating a Data Set for Each BY Group

The DATA step in Example 6.3 creates a data set for each BY group in a data set. The DATA step does not explicitly name the data sets it creates. Instead, statements determine the names of the data sets based on the BY values.

Example 6.3 uses three hash objects, one hash iterator object, the NUM_ITEMS attribute, and the CLEAR method to complete its task. It reuses the hash object that saves a BY-group. The CLEAR method empties the contents of this hash object before SAS reuses it.

Example 6.3 processes data set STUDYAPPTS. This data set contains height, weight, and blood pressure measurements for 24 subjects in a clinical study. The measurements were taken at five different times during the study: the start of the study, 30 days, 60 days, 120 days, and 180 days. Each observation in STUDYAPPTS contains the data for one subject's measurements at one of the five times.

Assume that the subjects started the study at different times and that the study has not yet finished. Therefore, a subject may have between 1 and 5 observations in STUDYAPPTS. All subjects in STUDYAPPTS have at least one measurement in addition to the baseline measurement.

The goal of Example 6.3 is to create a data set for each of the five measurement times, such that all observations for a measurement time are in one data set, and the data set's name reflects the measurement time. Within each time measurement output data set, the observations are arranged in descending order by the values of SYSTOL.

A limitation to Example 6.3 is that SAS derives the name of the data set from the BY-group value. When you adapt this program for your task, you will likely need additional statements that determine if the BY-group value is a valid SAS name. Further, if you have more than one BY-variable, your code should assign a name that reflects this situation.

Output 6.4 shows the data recorded for five subjects.

Output 6.4 PROC PRINT of STUDYAPPTS (first 5 subjects)

Obs	studyid	apptdate	appttype	weight	systol	diast
1	AFF115	03/07/2013	Baseline	175	153	88
2	AFF115	04/05/2013	One Month	174	151	84
3	AFF115	06/28/2013	Two Month	169	155	82
4	BBR123	03/08/2013	Baseline	163	155	93
5	BBR123	04/11/2013	One Month	158	162	91
6	BBR123	07/03/2013	Two Month	163	158	93
7	BBR123	11/01/2013	Four Month	168	161	93
8	BBR123	05/12/2014	Six Month	160	147	87
9	CDF108	02/04/2013	Baseline	244	143	89

Obs	studyid	apptdate	appttype	weight	systol	diast
10	CDF108	03/15/2013	One Month	252	147	91
11	CDF108	05/20/2013	Two Month	249	152	92
12	CDF108	10/11/2013	Four Month	256	152	94
13	DDL118	02/19/2013	Baseline	187	162	95
14	DDL118	03/28/2013	One Month	185	166	95
15	DDL118	06/07/2013	Two Month	181	167	93
16	DDL118	10/24/2013	Four Month	174	167	91
17	DDL118	05/22/2014	Six Month	170	157	84
18	DHZ122	03/08/2013	Baseline	138	142	83
19	DHZ122	04/18/2013	One Month	141	137	81

The first hash object, TYPES, that Example 6.3 defines contains the unique values of APPTTYPE. The values of variable APPTTYPE specify the measurement time. The NUM_ITEMS attribute applied to hash object TYPES obtains the number of unique values of APPTTYPE. This value equals the number of BY-groups in STUDYAPPTS.

SAS loads the contents of STUDYAPPTS into **the second hash object**, ALLOBS. Since the key item is APPTTYPE and the ORDERED: "YES" argument is on the DECLARE statement, SAS retrieves entries from ALLOBS in order by the values of APPTTYPE. With MULTIDATA: "YES" specified, SAS allows multiple data items per value of APPTYPE in hash object ALLOBS.

The third hash object, BYGRP, contains all the data for a single BY-group. The entries in BYGRP are arranged in order by the descending values of SYSTOL. The MULTIDATA: "YES" option ensures that multiple entries for the same value of SYSTOL in the BY group can exist in the hash object. Note that the DEFINEDATA method for BYGRP lists all of the data items. The ALL: "YES" argument tag can only be used when you specify a data set on DATASET: argument tag on the DECLARE statement. The ALL: "YES" argument tag cannot be used when the only way SAS fills the hash object data is as the DATA step executes.

Hash iterator object TI is associated with hash object TYPES. A call to the FIRST method finds the first value of APPTTYPE in TYPES. The iterative DO group executes once for each BY group since the upper bound is NTYPES. NTYPES is the variable whose value was determined by applying the **NUM_ITEMS attribute** to hash object TYPES.

For each BY group, the loop starts by applying the FIND method to find the first entry in ALLOBS with the current BY-group value that SAS retrieved from TYPES. The DO UNTIL loop then adds all the entries with the current BY-group value to hash object BYGRP, and it calls the FIND_NEXT method to find each of the remaining entries that have this current BY-group value. The DO UNTIL loop ends after SAS finds the last entry with the current BY-group value.

Next, the OUTPUT method outputs the contents of BYGRP to a data set whose name reflects the measurement time, which is stored in variable APPTTYPE. The COMPRESS function removes blanks from APPTTYPE so that the values “ONE MONTH”, “TWO MONTH”, “FOUR MONTH”, and “SIX MONTH” are changed to the valid SAS names of “ONEMONTH”, “TWOMONTH”, “FOURMONTH”, and “SIXMONTH”.

After the OUTPUT method executes, **the CLEAR method** removes the contents of BYGRP. The structure of BYGRP remains the same so it is not necessary to delete and redefine hash object BYGRP.

Example 6.3 Creating a Data Set for Each BY Group

```
data _null_;
  attrib studyid   length=$6
         appttype  length=$15
         apptdate  length=8 format=mmddyy10.
         weight    length=8
         systol    length=8
         diast     length=8;

  declare hash types(dataset: 'studyappts');
  declare hiter ti('types');
  types.definekey('appttype');
  types.definedata('appttype');
  types.definedone();
  ntypes=types.num_items;

  declare hash allobs(dataset: 'studyappts',multidata: 'yes',ordered: 'yes');
  allobs.definekey('appttype');
  allobs.definedata(all: 'yes');
  allobs.definedone();

  declare hash bygrp(multidata: 'yes',ordered: 'descending');
  bygrp.definekey('systol');

  bygrp.definedata('studyid','appttype','apptdate','weight',
                   'systol','diast');
  bygrp.definedone();

  call missing(appttype,studyid,apptdate,weight,systol,diast);

  rctypes=ti.first();
  do i=1 to ntypes;
    rcall=allobs.find();
    do until(rcall ne 0);
      rcadd=bygrp.add();
      rcall=allobs.find_next();
    end;
```

```
      rcout=bygrp.output(dataset: compress(appttype));
      rclr=bygrp.clear();
      rctypes=ti.next();
    end;
  run;
```

Example 6.3 creates five data sets: BASELINE, ONEMONTH, TWOMONTH, FOURMONTH, and SIXMONTH. Output 6.5 lists the observations in data set BASELINE.

Output 6.5 PROC PRINT of BASELINE

Obs	studyid	appttype	apptdate	weight	systol	diast
1	GHY101	Baseline	03/08/2013	180	180	90
2	REA102	Baseline	03/07/2013	154	173	85
3	DDL118	Baseline	02/19/2013	187	162	95
4	WCM111	Baseline	02/22/2013	140	156	90
5	BBR123	Baseline	03/08/2013	163	155	93
6	AFF115	Baseline	03/07/2013	175	153	88
7	TEP125	Baseline	03/01/2013	129	152	89
8	TRC107	Baseline	02/15/2013	135	152	85
9	YEA113	Baseline	02/28/2013	133	151	92
10	RGZ119	Baseline	02/22/2013	230	150	88
11	LOE121	Baseline	03/15/2013	255	149	82
12	NHC105	Baseline	03/15/2013	222	148	82
13	GAL114	Baseline	03/14/2013	144	146	82
14	CDF108	Baseline	02/04/2013	244	143	89
15	DRO116	Baseline	03/04/2013	168	143	79
16	TUY110	Baseline	02/08/2013	188	143	91
17	DHZ122	Baseline	03/08/2013	138	142	83
18	GEA117	Baseline	02/01/2013	199	140	86
19	UMB106	Baseline	03/11/2013	193	139	78
20	MBO120	Baseline	02/11/2013	203	138	85
21	VPD124	Baseline	02/12/2013	177	138	90
22	LIM109	Baseline	02/12/2013	173	137	76
23	MIJ104	Baseline	02/08/2013	205	135	90
24	NDM112	Baseline	02/13/2013	161	128	88

Output 6.6 lists the observations in data set SIXMONTH.

Output 6.6 PROC PRINT of SIXMONTH

Obs	studyid	appttype	apptdate	weight	systol	diast
1	GHY101	Six Month	06/05/2014	177	170	84
2	DDL118	Six Month	05/22/2014	170	157	84
3	TRC107	Six Month	05/22/2014	124	148	76
4	UMB106	Six Month	06/12/2014	187	148	64
5	WCM111	Six Month	04/29/2014	138	148	85
6	BBR123	Six Month	05/12/2014	160	147	87
7	GAL114	Six Month	06/09/2014	137	140	74
8	VPD124	Six Month	04/18/2014	190	137	86
9	TUY110	Six Month	12/25/2013	179	129	87
10	GEA117	Six Month	05/14/2014	172	127	86
11	DRO116	Six Month	05/26/2014	174	122	69

Comparing Two Hash Objects

Only one method exists that can compare two hash objects: the EQUALS method. This method evaluates whether two hash objects are identical. SAS considers two hash objects to be equal if the two hash objects have identical key and data item structure; the key values are equal; and the data item values are equal. Table 6.1 summarizes the rules that the EQUALS method applies when it compares two hash objects.

Table 6.1 Results of the EQUALS Method under Different Conditions

Key Items Are:	Key Values Are:	Data Items Are:	Data Item Values Are:	EQUALS Determines:
Same, and if both hash objects are ordered, the ORDERED: argument tag specification is the same.	Equal	Same	Equal	Equal (1)
Same, and if both hash objects are ordered, the ORDERED: argument tag specification is not the same.	Equal	Same	Equal	Unequal (0)
Same, and only one hash object is ordered.	Equal	Same	Equal	Equal (1)
Same	Unequal	Same	Equal	Unequal (0)
Different	Equal for the key items in common	Any condition	Any condition	Unequal (0)
Any condition	Any condition	Different	Equal for the data items in common	Unequal (0)

The syntax of the EQUALS method follows. SAS returns a 0 or 1 value to the variable named by the RESULT argument tag. The return code value reflects whether the EQUALS method successfully executed. Examine the value of the RESULT variable and not the method's return code when you work with the EQUALS method.

```
rc=object1.EQUALS(HASH: 'object2', RESULT: variable-name);
```

Example 6.4 compares four hash objects—NOTORDERED, ORDERED, ORDERED2, and ORDERED_D—where SAS loads the same data into each of the four hash objects, but the way they are defined is different. Example 6.4 processes data set CLAIMS2013. This data set contains several thousand observations of medical claims data. Each observation contains data for one medical claim for one patient.

Output 6.7 shows the first 20 observations in CLAIMS2013.

Output 6.7 PROC PRINT of CLAIMS2013 (first 20 observations)

Obs	ptid	claimdate	provider_id	charge	provider_type
1	4DGNWU3Z	04/12/2013	OPSURGERY039	$7,902.00	OPSURGERY
2	4DGNWU3Z	02/07/2013	PHYSTHERAPY071	$330.40	PHYSTHERAPY
3	XZRQTXGU	04/17/2013	OPRADIOLOGY069	$1,517.60	OPRADIOLOGY
4	JT1Z17GE	08/22/2013	CLINIC016	$53.00	CLINIC
5	THSALN2J	09/25/2013	OPRADIOLOGY070	$450.20	OPRADIOLOGY
6	THSALN2J	11/11/2013	CLINIC014	$229.80	CLINIC
7	THSALN2J	08/15/2013	CLINIC010	$102.20	CLINIC
8	THSALN2J	09/09/2013	LAB049	$147.40	LAB
9	DWJSEGFQ	07/31/2013	PHYSTHERAPY074	$381.40	PHYSTHERAPY
10	DWJSEGFQ	10/04/2013	CLINIC023	$65.80	CLINIC
11	L2PAT59T	10/16/2013	OPSURGERY044	$3,373.80	OPSURGERY
12	GE356WUG	10/07/2013	OPSURGERY035	$7,128.20	OPSURGERY
13	GE356WUG	07/24/2013	CLINIC005	$104.60	CLINIC
14	96O0JS7E	10/16/2013	CLINIC003	$280.40	CLINIC
15	96O0JS7E	04/08/2013	OPSURGERY036	$7,243.60	OPSURGERY
16	QMPH9WU3	11/04/2013	OPSURGERY041	$1,428.60	OPSURGERY
17	QMPH9WU3	07/03/2013	OPRADIOLOGY070	$3,180.20	OPRADIOLOGY
18	AD0ONT9C	03/21/2013	CLINIC012	$62.00	CLINIC
19	ZLEYIOE7	01/18/2013	PHYSTHERAPY074	$209.40	PHYSTHERAPY
20	ZLEYIOE7	08/20/2013	CLINIC002	$105.20	CLINIC

The only difference between hash object **NOTORDERED** and **ORDERED** is the presence of the ORDERED: "YES" argument tag in the definition for ORDERED. When SAS compares these two hash objects, it considers them equal even though the order in which key values are retrieved from each hash object is different. SAS assigns 1 to variable STATUS1.

The only difference between hash object **ORDERED** and **ORDERED2** is that ORDERED2 has an additional key item, PROVIDER_ID. When SAS compares these two hash objects, it considers them unequal, and it assigns 0 to variable STATUS2.

The only difference between hash object **ORDERED** and **ORDERED_D** is that the retrieval from ORDERED is in ascending key value order while retrieval from ORDERED_D is in descending key value order. When SAS compares these two hash objects, it considers them unequal, and it assigns 0 to variable STATUS3.

Example 6.4 Determining If Two Hash Objects Are Equal

```
data _null_;
  attrib ptid length=$8 label='Patient ID'
         claimdate length=8 format=mmddyy10. label='Claim Date'
         provider_id length=$15 label='Provider ID'
         charge length=8 format=dollar10.2 label='Charge'
         provider_type length=$11 label='Provider Type';

  array yesno{0:1} $ 6 _temporary_ ('0: No','1: Yes');

  declare hash notordered(dataset: 'claims2013');
  notordered.definekey('ptid','claimdate');

  notordered.definedata('provider_id','charge','provider_type','claimdate',
                        'ptid');
  notordered.definedone();

  declare hash ordered(dataset: 'claims2013',ordered: 'yes');
  ordered.definekey('ptid','claimdate');
  ordered.definedata('provider_id','charge','provider_type','claimdate',
                     'ptid');
  ordered.definedone();

  declare hash ordered2(dataset: 'claims2013',ordered: 'yes');
  ordered2.definekey('ptid','claimdate','provider_id');
  ordered2.definedata('provider_id','charge','provider_type','claimdate',
                      'ptid');
  ordered2.definedone();

  declare hash ordered_d(dataset: 'claims2013',ordered: 'd');
  ordered_d.definekey('ptid','claimdate');

  ordered_d.definedata('provider_id','charge','provider_type','claimdate',
                       'ptid');
  ordered_d.definedone();

  call missing(provider_id,charge,provider_type,claimdate,ptid);

  rc=notordered.equals(hash: 'ordered',result: status1);
  put "NOTORDERED=ORDERED: " yesno{status1};

  rc=ordered.equals(hash: 'ordered2',result: status2);
  put "ORDERED=ORDERED2: " yesno{status2};

  rc=ordered.equals(hash: 'ordered_d',result: status3);
  put "ORDERED=ORDERED_D: " yesno{status3};
run;
```

The SAS log that shows the output from the three PUT statements follows.

```
NOTORDERED=ORDERED: 1: Yes

ORDERED=ORDERED2: 0: No

ORDERED=ORDERED_D: 0: No
```

Specifying Memory Structure Usage of a Hash Object

As your applications become more complex, and if the sizes of your hash objects are large, you might find that you can improve processing times if you specify the HASHEXP: *n* argument tag on the DECLARE statement. The HASHEXP: *n* argument tag provides some user control over how SAS structures your hash object. Determining the best value for *n* involves trial and error and examination of the processing statistics in the SAS log.

The value that you supply to HASHEXP: is used as a power-of-two exponent. This 2^n value is not the number of items in your hash object. Instead the 2^n value is the number of "buckets" you want SAS to allocate to manage the lookup process in working with your hash object. SAS does not have a maximum number of items it can put in each bucket. It has its own set of algorithms that determine how much to place in each bucket. When you specify HASHEXP: 4, you tell SAS to create 2^4=16 buckets.

In general, the bigger your hash object, the more likely that your DATA step will execute faster if the value you assign to HASHEXP is greater. The default value assigned to HASHEXP is 8. All examples in this book use this default setting.

An example of a DECLARE statement that specifies the HASHEXP: argument tag follows.

```
declare hash c(dataset: 'claims2013', hashexp: 12);
```

Determining the Size of an Entry in a Hash Object

When you want to better understand how SAS processes your hash object, you might want to find out the size of an entry in your hash object. With this information, you can better judge whether you need to modify the structure of your hash object and your DATA step so that memory is more optimally used.

You can apply the ITEM_SIZE attribute to a hash object to find the size in bytes of an *entry* in a hash object. The value that it returns is the number of bytes to hold a key item, its associated data items, and internal information that SAS maintains.

A formula does not exist that can connect the value that ITEM_SIZE returns with the value that you assign to the HASHEXP: argument tag. Through trial and error adjustment of the value you supply to HASHEXP and the structure of the associated hash object, you can determine if changes to HASHEXP and your hash object make a significant difference in the processing of your DATA step. (As of SAS 9.3, SAS online sample 34193 supplies a macro that estimates the amount of memory a hash object takes.)

Example 6.5 calculates the item size for three hash objects. SAS loads the same key and data items into each of the three hash objects, but the hash objects have different definitions. Example 6.5 processes data set OCTOBEREVENT, which contains the schedule information for 52 employees who are assigned to work at a health screening event on October 22 and 23, 2013. An employee can have multiple observations in OCTOBEREVENT.

Output 6.8 lists the first 20 observations in OCTOBERVENT.

Output 6.8 PROC PRINT of OCTOBERVENT (first 20 observations)

Obs	eventdate	shift	activity	empid	empname
1	10/22/2013	AM	Cholesterol	02QYJG	Howard, Rachel Y.
2	10/22/2013	AM	Cholesterol	8I4YKY	Brooks, Shelby B.
3	10/22/2013	AM	Cholesterol	ADHW3A	Moore, Allison W.
4	10/22/2013	AM	Dietary	IFQZ8S	Nelson, Kelly U.
5	10/22/2013	AM	Dietary	OYMEE3	Thomas, Hannah I.
6	10/22/2013	AM	Dietary	PTBHUP	Thompson, Olivia P.
7	10/22/2013	AM	Pharmaceutical	WVV7PT	Patterson, Vanessa N.
8	10/22/2013	AM	Counseling	14ZN75	Miller, Sierra Q.
9	10/22/2013	AM	Counseling	5KA7JH	Martin, Aaron G.
10	10/22/2013	AM	Counseling	BA8CRZ	Baker, Marissa R.
11	10/22/2013	AM	Counseling	FAL5UZ	Butler, Haley P.
12	10/22/2013	AM	Counseling	TMJTVP	Edwards, Brittany O.
13	10/22/2013	AM	Counseling	YXW78P	Davis, Andrea Y.
14	10/22/2013	AM	Immunization	1CV01H	Brown, Lindsey V.
15	10/22/2013	AM	Immunization	8EHCSX	Hernandez, Elizabeth S.
16	10/22/2013	AM	Immunization	HD8ERT	Paine, Mike J.
17	10/22/2013	AM	Immunization	KRTV2T	Gonzalez, Alejandro F.
18	10/22/2013	AM	Vitals	NCPO4E	Richardson, Bailey U.
19	10/22/2013	AM	Pharmaceutical	P438U8	Price, Katelyn Q.
20	10/22/2013	AM	Vitals	2S57ZI	Thompson, Alexandria B.

The differences among the three hash objects in Example 6.5 are:

- Hash object OEUNIQUE does not allow multiple sets of data items per key value. With omission of the ORDERED: argument tag, SAS does not retrieve entries from OEUNIQUE in any specific order.
- Hash object OEMULT allows multiple sets of data items per key value. With omission of the ORDERED: argument tag, SAS does not retrieve entries from OEMULT in any specific order.
- Hash object OEMULTORDER allows multiple sets of data items per key value. With the ORDERED: argument tag included, SAS retrieves entries in order by the values of EMPID.

Example 6.5 Determining Sizes of Hash Objects

```
data _null_;
  attrib eventdate format=mmddyy10.
         shift length=$2 label='Event Shift'
         activity length=$20 label='Event Activity'
         empid length=$6 label='Employee ID'
         empname length=$40 label='Employee Name';

  declare hash oeunique(dataset: 'octoberevent');
  oeunique.definekey('empid');
  oeunique.definedata('eventdate','shift','activity','empid','empname');
  oeunique.definedone();
  declare hash oemult(dataset: 'octoberevent',multidata: 'yes');
  oemult.definekey('empid');
  oemult.definedata('eventdate','shift','activity','empid','empname');
  oemult.definedone();

  declare hash oemultorder(dataset: 'octoberevent',multidata: 'yes',
                           ordered: 'yes');
  oemultorder.definekey('empid');
  oemultorder.definedata('eventdate','shift','activity','empid','empname');
  oemultorder.definedone();

  call missing(eventdate,shift,activity,empid,empname);

  uniquesize=oeunique.item_size;
  multsize=oemult.item_size;
  ordersize=oemultorder.item_size;

  put "ITEM_SIZE for OEUNIQUE=" uniquesize;
  put "ITEM_SIZE for OEMULT=" multsize;
  put "ITEM_SIZE for OEMULTORDER=" ordersize;
run;
```

PUT statements write in the SAS log the values that the three calls to ITEM_SIZE return.

```
ITEM_SIZE for OEUNIQUE=104

ITEM_SIZE for OEMULT=112

ITEM_SIZE for OEMULTORDER=112
```

If you sum the number of bytes that the items take, the sum is 76 bytes, which is less than any of the values that ITEM_SIZE returns. SAS adds its own internal information to each entry to manage the hash object. In Example 6.5, it can be observed that the MULTIDATA: “YES” argument tag adds to the size of an entry while the ORDERED: “YES” does not.

References

Hash object programming is a popular topic for SAS user group meetings and on the SAS Community Web site. As of the printing of this book, a search for "hash object" on http://support.sas.com yielded 66 technical papers, 80 SAS Global Forum papers, and 48 samples and SAS notes.

Hash object programming is also discussed in these SAS Press books:

Burlew, Michele M. 2009 *Combining and Modifying SAS Data Sets: Examples, Second Edition*, Cary, NC: SAS Institute, Inc.

Carpenter, Art. 2012. *Carpenter's Guide to Innovative SAS Techniques*, Cary, NC: SAS Institute, Inc.

SAS Institute Inc., 2011. *SAS 9.3 Component Objects: Reference*, Cary, NC: SAS Institute, Inc.

Index

A

B

C

D

E

F

I

K

L

M

N

O

W

CPSIA information can be obtained at www.ICGtesting.com
Printed in the USA
LVOW11s1124150813

347751LV00003B/25/P